GEOGRAPHIC INFORMATION SYSTEMS

TOR BERNHARDSEN

ISBN 82 – 991928 – 3 – 8

Published by:
Viak IT
Longum Park
P.O.Box 1699 Myrene
4801 Arendal
Norway

Printed by:
Norwegian Mapping Authority

Layout and illustrations:
AB Trykk, Arendal

INTRODUCTION

As this is written the people of the world find themselves at a major turning point in the history of this century and possibly in the history of civilization on earth.

The end of seventy years of soviet communism, the disintegration of the former U.S.S.R., and the resulting thaw in East-West relations have created a moment almost unique in this century and mark an event which many of us did not expect to see in our lifetimes: the apparent end of more than forty years of the «cold war».

These events have created an opportunity unique in this century for the nations of the world to cooperate in dealing with the many problems which face the global community at this time. The attendance of the heads of state of more than one hundred countries at the Earth Summit in Rio de Janeiro in June of 1992 symbolizes the nature of the cooperative efforts which the members of the United Nations are making to solve their common problems and create a better common future.

In approaching global problems, the people of the world are also benefiting from fifty years of major developments in the science and technology used for studying the earth, its human population, and the natural and cultural resources which make up the global environment. Especially important among these have been the rapid developments in remote sensing of the earth environment, in computer science and technology, in global electronic communications, and in the gathering of information about the earth and its inhabitants into vast new electronic storehouses.

In sum, as we enter this last decade of the twentieth century, we are in the fortunate position of having both the technical and the political means of bringing the people of the world together to deal with the world's problems: we must seize this opportunity.

One of these newly-developed technologies and one of the most important components of any approach to global problem solving is Geographic Information Systems (GIS) technology. Developed in just the last 25 years or so, GIS technology already represents a billion dollar industry world-wide, growing at perhaps 25% per year, and serving some 50,000-100,000 users in more than 70 countries. If projections are correct, GIS will be used by more than a million persons by about the turn of the century.

GIS has been used on problems at a wide range of scales and for geographic areas from a few hectares up to the global databases just now beginning to become available for wide use. GIS has been applied by a wide range of disciplines to a correspondingly wide range of problems. Governments, non-governmental organizations, businesses and educational institutions all now use GIS technology. Members of the general public will become GIS users in the years just ahead. That GIS technology has proven its value to its users is indicated by the rapid growth in its use, the rapid growth in expenditures for GIS technology, and by the very large amounts of resources now being devoted throughout the world to creating digital data for use in GIS of various kinds.

Another indication of the value of GIS technology is that the number of GIS educational programs is growing even faster than is the use of the technology. For this important educational enterprise to succeed, sound textbooks in GIS technology are needed. This volume is just such a textbook. Until just a few years ago there were virtually no introductory textbooks dealing with GIS technology. Those who learned the technology learned chiefly or exclusively through experience. Fortunately that is now changing as new GIS textbooks are published nearly every month. Nevertheless, because the field is so new, the contents appropriate to such a textbook are not yet widely agreed upon. To a degree then, each new textbook is still an experiment, a creation not unlike those experimental vessels which vikings of an earlier age built so assiduously and successfully, and by means of which they extended themselves over the seas of much of the known world. I think this book is also well-designed, solidly constructed, and finely crafted; those who depend on it as they set out to explore our spatial world will be well served.

In my own work in the GIS field over the last twenty-five years or so, I have become convinced of the importance of providing GIS technology which works for its users. Many persons, especially those new to the field, can be overwhelmed by the rapid growth of the GIS literature and by the hyperbole which is often associated with the announcement of new products and new developments, many of which will not advance the field and some proportion of which will never, in fact, actually come into existence. Students and others new to the GIS field need the guidance of textbook authors with extensive experience in the field, discernment, and careful judgement; authors who can separate the sound and essential from that which will be ephemeral; authors who can indicate what actually works. That experience is reflected in these pages.

It's also important to realize that a GIS is considerably more than just technology; most of the problems with real GIS now have more to do with the people and the procedures they use (system design, applications programming, effective use of GIS in solving real problems, funding new systems, and so on) than with computer hardware or software. The reality is that while one can now buy a reliable turnkey GIS hardware/software system and can often obtain a great deal of GIS data through purchase or conversion, there is no similarly reliable way to create the organizational structures and experienced staff necessary to support an effective GIS. These things must be learned, and, unfortunately, their important nuances probably cannot be taught.

Some years ago, in thinking about how GIS use could best be developed throughout the world, I made a list of the traits that I thought GIS technical advisers would require if they were to be successful in spreading GIS use into new disciplines, new organizations, and new areas in the world. The list of intellectual and personal traits was a formidable one, chiefly because I believe GIS technology can be applied to so many kinds of problems and in so many situations. The list, in turn, led me to reflect on how one could possibly find, recruit, and train such people. How does one go about producing people with boundless enthusiasm, extraordinary patience, coupled with appropriate technical, managerial, and political skills; wholists who are technicians as well as humanitarians; persons sympathetic to the ills of humankind and not afraid to wade in and try to improve things even when, often, they will not be successful?

In the end, I have concluded that such persons will, naturally and inevitably, be drawn to GIS technology, bringing with them the many skills and abilities they have already acquired elsewhere. When they understand just what GIS technology is capable of doing for many kinds of problem solving, they will embrace it and take it with them wherever they go. Books like this one will foster that process, and so contribute to the improvement of the world in which we live.

I hope that the readers of this textbook will profit from what it has to teach them. If we are to solve many of the problems facing us -- in the cities, in the wild areas of the earth, in the atmosphere and the oceans, problems of the earth as a whole-- we shall need the help of skilled users of GIS technology. If readers can master what is in this volume, they will be well started on this enterprise.

Jack Dangermond
May, 1992

GEOGRAPHIC INFORMATION SYSTEMS

FOREWORD

This book is an attempt to meet the need for a comprehensive presentation of the various fields currently associated with the term «geographic information systems» (GIS). Very few limitations have here been set on interpretations of the term GIS. By looking at the table of contents, one might find that the book could just as well have been entitled «Geographic Information Technology» (GIT). In allowing for a wide range of definitions we have taken into consideration the fact that most users of geographic data are addressing specific tasks, and that they will always choose the technology most applicable to those tasks, regardless of whether or not the technology falls within a particular definition of GIS.

Questions related to the introduction of GIS into organizations are in this book treated in relatively great depth. This is because correctly addressing these issues is directly linked with the successful functioning of the systems being introduced. This is the reasoning behind the sub-title «the beneficial use of geographic information technology».

The idea for this book arose out of my work with the Nordic joint project named «Community Benefits of Digital Spatial Information – Nordic KVANTIF I & II», 1985-1990.

The first edition of the book came out in Norwegian in 1989. During this first preparation I was given invaluable assistance, in particular by Professor Øystein Andersen at the Agricultural University of Norway, Harald Danielsen of the Nidar District Administration, and Arild Reite of the Norwegian Cartographic Union.

Prior to this second edition, which is now being published in English, an extensive revision has been carried out. With regard to this I would like to express my particular appreciation of the assistance given again by Professor Øystein Andersen and by the Director of GRID-Arendal, Olav Hesjedal, both of whom have provided valuable suggestions as to how the book might be improved. Svein Tveitdal, the International Director of Viak I.T., has also made important contributions to Chapter 12. Knut Heggen of Asplan Viak Sør has provided practical examples for Chapter 9, and Robert Sandvik of the Norwegian Naval Cartography Administration has supplied material for Chapter 13.

I would like to express my gratitude for the fine cooperation and teamwork of Michael Brady Consultants who have been responsible for the

VIII

English translation. They have also made contributions in other respects, thereby improving the overall quality of the book. Dr. William Warner at the Agricultural University of Norway has made a major contribution by reviewing the technical and scientific terminology of this English edition. I would also like to thank Jürg Keller of AB-Trykk for the fine layout and illustrations.

Finally, I would like to express my deepest thanks to my wife and two children for their patience and forbearance while so much of my free time has been spent working on the book.

Arendal, May 1992

Tor Bernhardsen

GEOGRAPHIC INFORMATION SYSTEMS

13. ELECTRONIC CHART DISPLAY AND INFORMATION SYSTEM (ECDIS)

1   GEOGRAPHIC INFORMATION SYSTEMS AND GEOGRAPHICAL INFORMATION

1.1   BASIC CONCEPTS

Computerisation has opened new vistas in both documentation and the prognoses essential to all decision-making and dissemination of information. Data representing the real world can be stored and processed so that it can be presented later in simplified forms to suit specific needs. As illustrated in Fig. 1.1, when the manipulation and presentation of data relates to geographical locations, our understanding of the real world is enhanced.

Since the mid 1970s, specialised computer systems have been developed to process georeferenced information in various ways. These include:

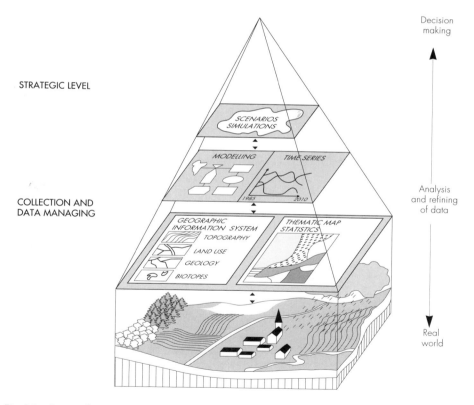

Fig. 1.1 – By use of geographic information systems, a "simplified" world can be brought into the computer – different techniques can be applied to analyse and "simplify" the data, and the foundation is laid for the "decision-making pyramid". Today, geographical information systems are in the process of filling the upper half of the pyramid. (figure freely adapted from Grossman, 1983).

- organising parts of the information available
- locating specific information
- executing computations, illustrating connections and performing analyses that were previously impossible.

The collective name for such systems is Geographic Information Systems, abbreviated GIS. Geographic Information Systems may be viewed as the result of a marriage of Computer Assisted Cartography (CAC) and database technology.

Traditionally, geographic data is presented on maps using symbols, lines and colours. Most maps have a legend in which these geometric elements are listed and explained – a thick black line for main roads, a thin black line for other roads, and so on. Dissimilar data can be superimposed on a common co-ordinate system. Consequently, a map is both an effective medium for presentation as well as a bank for storing geographic data. But herein lies a limitation. The stored information is processed and presented in a particular way, and usually for a particular purpose. Altering the presentation is seldom easy. A map provides a fixed, static picture of geography that is almost always a compromise between many differing user needs.

Maps are a considerable public asset. Surveys conducted in Norway indicate that the aggregate benefit accrued from using maps is three times the total cost of their production.

Compared to maps, GIS has the inherent advantage that data storage and data presentation are separate. As a result, data may be presented and viewed in various ways.

Ever since the earliest prototype systems of the 1970s were field tested, GIS has grown to include numerous applications in the public and private sectors. The growth potential is enormous: the worldwide hardware and software market for GIS between 1990-1995 is estimated at 12 billion US dollars. Annual growth in the early 1990s was 30%. (Dataquest, 1991).

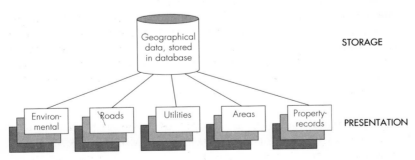

Fig. 1.2 – A map can be a combined presentation and storage medium, with resultant limitations. With GIS, storage and presentation are separated, thereby enabling a wide variety of products to be created from the same basic data.

GIS defined   Definition of GIS

Although there is no universal definition for computer systems manipula-ting geographic data, Geographic Information Systems (GIS) is the collective term commonly accepted for them. These systems are implemented with computer hardware and software functions for the:

- acquisition and verification
- compilation
- storage
- updating and changing
- management and exchange
- manipulation
- retrieval and presentation
- analysis and combination

of geographic data, which may be defined as consisting of "information on the qualities of and the relationships between objects which are uniquely georeferenced."

The qualities involved may be physical parameters such as length, area, weight and temperature, as well as classifications according to the type of vegetation, historic group, zoning etc. Such occurrences as accidents, floods and alterations may also be included.

The relationships between geographic entities often provide vital infor-mation. For example, the details of a water supply pipe network may be critical for firemen who need to know which valves to close in order to increase the water supply pressure whilst extinguishing a fire. The details of properties bordering a road are necessary if all property owners affected by road works are to be properly notified.

Georeferencing is usually expressed in terms of positions in Cartesian coordinates (such as Northing, Easting, Elevation) or in latitude and longitude. But other reference systems such as postal codes and various area divisions used in map indexes and demographic studies are also employed.

Stored data may be processed in a GIS for presentation in the form of maps, tables or special formats. One major GIS strength is that dissimilar data can be linked. Thus, in a given area the interests of conservationists and developers can be compared or health demography and environmental impact correlated. The only common denominator required is that the various areal dif-ferentiations are expressed in compatible reference systems.

A Geographical Information System can process georeferenced data and provide the answers to questions involving, say, the particulars of a given loca-tion, the distribution of selected phenomena, the changes that have occurred since a previous viewing, the impact of a specific event, or the relationships and systematic patterns of a region.

DIGITAL MAP DATABASE   COMPUTERISED TABULAR DATA

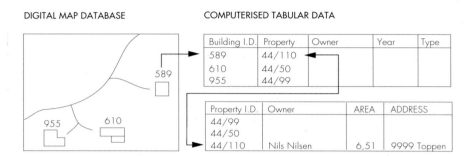

Fig. 1.3 – This figure illustrates how GIS functions - based on the interaction between a digital map database and computerised tabular data.

As often emphasised, GIS can perform spatial analyses of georeferenced data to illuminate such specifics as the quickest driving route between two points, or the dependence of contested regional planning areas on the varying weightings of conflict parameters (such as preservation vs. development).

Many modern GIS can process data from dissimilar sources, including digital map data, digital images, video images, computer-aided design (CAD) data and various computer-based registers. Consequently, a GIS might be termed a "data mixing system."

Technically, GIS organises and exploits digital map data stored in databases. In GIS, the real world is described using digital map data, which define positions in space, and attribute data, which usually consist of alphanumeric lists of characteristics and, frequently, temporal information describing when the other data are valid in time. Logical links between geometric entities may also be described, including road connections, common property boundaries and the like. Attribute data may be converted to graphic symbols presented together with other data on a map. Conversely, the GIS may be implemented so that attribute data can be retrieved merely by moving a pointer to a symbol on a screen display and keying in a command. In GIS, geometric data and attribute data are usually separated, in software hierarchy. Identical identifiers in the two databases facilitate matching for retrieval and processing.

Stored GIS data do not resemble the visual presentations of traditional maps. GIS data are usually stored in databases in ways that permit new insights, particularly into the relationships between dissimilar entities. Therefore, the types, positions, and interrelationships of all entities must be known before relationships are determined.

Databases are vital in all Geographic Information Systems. A database is a comprehensive collection of related data stored in logical files and collectively processed, usually in tabular form. A Database Management System (DBMS) is essentially software that manipulates, i.e. imports, stores, sorts and retrieves, data in a database.

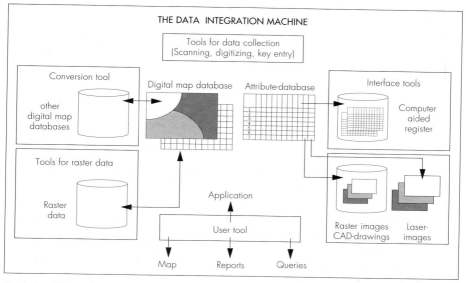

Fig. 1.4 – GIS is in the process of becoming a typical "data integration machine". Some modern systems can receive, process and transmit data of widely varying origin. (figure freely adapted from ESRI)

Data in a database:
- are stored and maintained in one place
- stored in a uniform, structured, and controlled manner
- can be divided as needed, and
- are easily updated with new data.

This contrasts with manual systems, in which data often:
- are stored by many different agencies,
- in various files and reports of varying accessibility,
- on maps with scales coordinates so diverse that they cannot be collocated
- usually difficult to update.

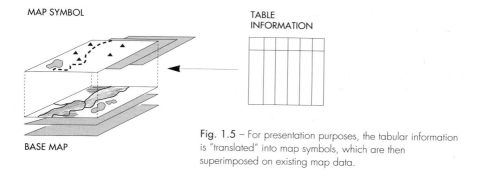

Fig. 1.5 – For presentation purposes, the tabular information is "translated" into map symbols, which are then superimposed on existing map data.

The constituent parts and modes of operation of GIS are discussed in detail in Chapter 3.

GIS diversity

Whilst this general definition of GIS is quite valid, in practice the diversity of GIS has spawned various definitions. Firstly, users have contrived working definitions suited to their own specific uses. Thus, they may vary according to whether operators are planners, water supply and sewage engineers, or support service personnel – or perhaps public administrators. Secondly, those with a more theoretical approach – research workers, software developers or sales and training staffs – may use definitions deviating from the purely practical.

Furthermore, systems may be seen as tools tailor-made from selections of semi-independent software modules, computer hardware and other devices. Yet a GIS that functions as a map-making machine can also be used as an analytical tool.

Depending on its use, a GIS may be viewed as a pure:

- Data processing system, e.g for: – map production
  – 3D visualisation
- Data analysis system, e.g. for: – conflict analyses
  – optimising transport
- Information system, e.g. for: – public services
  – documentation and case work
- Management system, e.g. for: – network operation and
  maintenance
  – real estate management
  – natural resources management

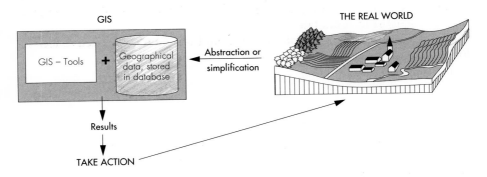

Fig. 1.6 – GIS is intended to be a means of improving man's everyday life. It is therefore important that the information which results from data processing be applied to guide "the real world" in the right direction.

- Planning system, e.g. for:
  - roads and other transport systemes
  - excavations, canals, etc.
  - water supply and sewage forest harvest operation
- Electronic navigation system

Characteristically, the field of GIS involves many disciplines, uses, types of data and end users:

- disciplines: computer sciences, cartography, photogrammetry, surveying, remote sensing, geography, hydrography, statistics, organisational development, planning, etc.
- uses: operation and maintenance of networks and other facilities, management of natural resources, real estate management, road planning, map production, etc.
- data: digital maps, digital imaging of scanned maps and photos, satellite data, ground truth data, video images, tabular data, text data, etc.
- users: water supply and sewage engineers, planners, biologists and geologists, politicians and other decision makers, cartographers, surveyors, etc.

Geographic Information Systems are not yet available off-the-shelf; only their constituent devices, such as computer hardware and basic software can be bought. So a GIS can function only after the requisite expertise is available, the data are compiled, the various routines organized and the programs modified. These facets of an overall GIS are interlinked, as illustrated in Fig. 1.7.

In general, procurement of the various devices, and computer hardware and software, is vital but straightforward. The expertise required is often underestimated, the compilation of data is expensive and time consuming, and the organisational problems are the most vexing. These facets of an overall GIS will be discussed in detail later.

GIS developments and applications should focus on cost-effective technologies tailored to user needs.

GIS – CHAIN

Fig. 1.7. – A GIS system cannot be bough "off the shelf". The system has to be built up within an organisation. When planning to introduce GIS, it is important that equal attention be given to all four links in the GIS chain.

The classification of any given technology with respect to a particular definition of GIS is of minor interest. Therefore, a broad definition of GIS best serves the interests of most users. Geographic Information Technology (GIT) might then be the better term.

In its infancy, GIS in many ways resembled Computer Aided Design (CAD). Both focused on graphic images. Therefore, the early GIS systems were modified versions of CAD systems. By the early 1980s, however, GIS had become distinctly different from CAD and systems were developed specifically for GIS applications. Fundamental differences underlie this trend. Foremost, GIS analyses data, while CAD constructs and illustrates data. A GIS can take two or more data sets, analyse their integration, and generate a totally new data set; CAD cannot. GIS applications, for example, are more numerous than those of CAD. They are also different. For instance, non-graphic GIS information can be more important than graphic presentation, as in administrative applications for municipal planning, real estate information, network age, etc. From a pure data manipulation viewpoint, GIS systems differ because they process far more data than most other comparable computer systems. As discussed later, this characteristic sets special requirements for database structure and memory capacity.

GIS systems are often designated according to application. Hence the titles Land Information System (LIS), Urban Information System and Natural Resource Information System. In international publications, the term spatial data is often used in place of geographic. Automatic Mapping and Facility Management (AM/FM) is frequently used for systems dedicated to the operation and maintenance of networks. Therefore AM/FM is now most often used by public utilities and transportation systems. Nonetheless, GIS is now internationally the most accepted term and has the advantage that it can be applied to a spectrum of systems that process geographic data.

1.2   SOCIO-ECONOMIC CHALLENGES

According to the Nordic KVANTIF and other surveys, between 50%-70% of the data involved in local administration are geographic. These surveys have also shown that major users of cartographic and geographic data – the construction sector, public administration, agriculture, forestry and other resource management, telecommunications, electricity supply, transportation, etc. – spend 1.5% to 2% of their annual budgets on cartographic and geographic data. In relation to gross national product (GNP), annual expenditures average 0.5% in industrialised countries and 0.1% in developing countries. The figures for industrialised countries are not expected to increase, but GIS is expected to account for an increasingly greater share of the overall annual figures – particularly in sectors where traditional maps and geographic data have been

dominant but have now been found lacking. The figures for developing countries are expected to increase in pace with development.

By the early 1990s, technical problems had become insignificant. The major
challenges to system developers and users alike are filling the gaps between
available technology and practical application.

The future importance of GIS is contingent upon systems being developed
to meet concrete needs. There are several sectors in which GIS has the potential
to play a major role.

The complex society

Modern societies are now so complex, and their problems so interwoven,
that they cannot be solved independently. For instance, a new housing development may affect the local school system. Altered age distribution in a
village may affect health and social expenditure. The volume of city traffic
volume may put constraints on the maintenance of buried pipe networks,
affecting health. Street excavations may drastically reduce the turnover of retail
shops. Traffic noise from new roads or motorways may well force people from
their homes.

The actions needed to solve such problems are best taken on the basis of
standardised information which can be combined in many ways to serve many
users. Geographic Information Systems have this capability.

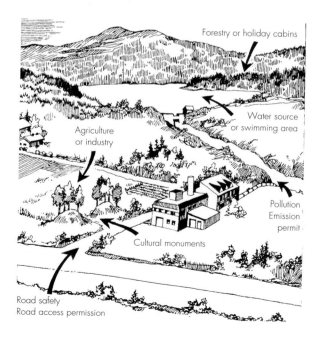

Fig. 1.8 – Today's society has a
growing need for information.
Conflicts of interest have to be
brought out into the open and
information has to be made
available to the concerned
parties. It has been shown that at
least 50% of the public
administration sector needs for
geographical data in one form or
another. (The Norwegian
Association for Cartography
et.al.).

Populations are now more mobile than ever; changing jobs and moving house have become commonplace. When key personnel leave, they take their expertise with them. If that expertise involves specific knowledge of, say, the water supply and sewage network of a community, the loss can be serious if the information is otherwise inadequately documented. Here, too, GIS has an advantage in that it can act as an effective filing system for dissimilar sectors of a complex society.

Operation and maintenance

Another aspect of developed industrial societies is that emphasis has shifted from planning and development to operation and maintenance, particularly for such works as buried pipes and cables.

Assuming that the costs of municipal operation and maintenance of water supply and sewage networks are proportional to the aggregate pipe length, overall expenditures can be expected to increase as pipes get older. As shown in Fig. 1.9, the aggregate pipe length in Sweden doubled between the mid-1960s and the mid-1970s. The result: operation and maintenance costs in the 1990s are double those of the 1980s.

In 1990 in Norway, aggregate municipal spending for operation and maintenance was nine times that allocated to building new facilities.

The large financial sums involved indicate that even minor gains in efficiency in this area can result in considerable savings. Yet, more efficient operation and maintenance is possible only when full information is available on the facilities' location, condition, age, construction material(s), service record, etc.

As many as 35 different types of piping and cable may be buried beneath the surface of a major city street. These include telecommunications and optical cables, high-voltage, local power, signal, heating and TV cables, as well as piping for water supplies, sewerage, gas and remote heating. Indeed, such is the confusion of pipes and cables that it is virtually impossible to include all the

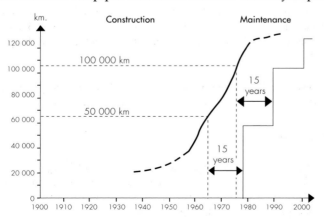

Fig. 1.9 – This graph shows the total length in kilometres of the water and sewer systems in Sweden from 1935 to date. It also shows that maintenance is normally carried out about 15 years after the pipes have been laid. This means that maintenance costs in the 1990's will be double those of the 1980's.

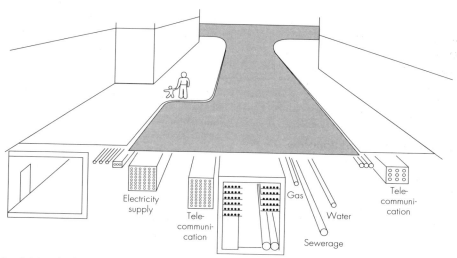

Electricity
supply

Tele-
communi-
cation

Gas

Water

Tele-
communi-
cation

Sewerage

Fig. 1.10 – This figure illustrates the complexity of the systems to be found under a city street. There can be up to 35 different utility systems under one single street.

relevant information on an ordinary map. Therefore, starting in the mid 1970s, piping and cable data have increasingly been computerised.

Computerisation is the only realistic way of systematising and standard-ising the enormous amounts of data involved. For instance, even in a smaller city such as Oslo, Norway (pop. 500,000) there are five million meters of buried piping and cables, registered on 20 different types of maps totalling some 4,500 map sheets.

Even though the cities of industrialised countries have long had such problems, the overall picture seems equally serious in the cities of developing countries. In some cases, construction workers have been electrocuted when high-voltage cables whose locations were not well known have accidentally been cut. The situation with water supply and sewage networks is equally adverse. Countless human tragedies may be ascribed to inadequate, leaky water supply, deplorable sewerage, lack of maintenance and lack of knowledge of the networks involved.

The prime requirement for the information involved is, of course, that it must be manageable. This means that in the future it must be computer-based.

Environmental and reource management

Decision-making is becoming increasingly complex as dwindling natural resources and more demanding economic priorities diminish the chances of today's decision being right tomorrow. Furthermore, environmental awareness is ever increasing. The pressing global challenges are now:

Fig. 1.11 – Today the world's natural resources are under great pressure and it is important to ensure a sustainable development built on a strong foundation. GIS can be instrumental in initiating such a process (I. J. Jahnsen).

- uncontrolled desertification
- erosion
- monoculture and soil depletion
- dissemination of endangered animal and plant species
- acid rain and fish deaths
- pollution of rivers, lakes, and oceans
- dying forests
- environmentally-related illnesses
- contaminated ground water and other environmentally-related strains
- greenhouse effect and resultant climatic change
- reduced ozone layer
- repeated environmental catastrophes (oil spills, poison leaks, radioactive material spreading, etc.)

Despite comprehensive and lengthy studies, the global environment is still not well understood. This is because nature is complex and most effects are inter-related. For instance, a small decrease in the atmospheric ozone layer permits more ultraviolet radiation to reach the surface of the Earth, which kills marine algae on which fish feed. This, in turn, disseminates fish, which reduces catches, possibly threatening the livelihood of fishing villages in major coastal areas. (GRID, 1991)

Therefore, many countries now have programmes to register existing natural resources and known sources of pollution. The United Nations Environmental Programme is also involved and is currently establishing a global network of environmental databases.

Environmental data may be used both to expose conflicts and examine environmental impacts. Impact analyses and simulated alternatives will probably become increasingly important. GIS may play a key role in:

- documenting natural conditions and developments
- documenting the suitability of resources for various uses
- exposing conflicts/conflicting interests
- revealing cause-effect relationships

Addressing environmental problems is complicated by limited information. Even the relatively sparse data that have been compiled are of limited use in decision-making because of non-uniform storage/filing, lack of verification and a lack of system or hierarchy in data acquisition, obsolescence, etc. GIS techniques using databases seem ideally suited for manipulation such environmental data.

Planning and development

As discussed above, the planning and development of housing, roads, industrial facilities and the like requires data on the terrain and other geographic information.

Development often involves building on more difficult or rugged terrain than before, or increasing the density of building in areas already built-up, or both. Yet, the new structures must fit within the existing technical infra-structure and computerisation is therefore a great aid. For instance, when the Stockholm's Central Railway Station was expanded between 1986 and 1989, all

Fig. 1.12 – Today's requirements for physical planning are stringent. Projects which involve encroachments on nature are required to be carried out in the most environmentally-friendly but economical way (VIAK System).

technical infrastructure facilities above and below ground were registered, and the entire project managed with the use of computer and GIS systems (see Chapter 14. One of the benefits realised was that there were very few interruptions to railway operations and road/street traffic.

Escalating construction costs have made optimum building and road location extremely important. Minimising blasting and earth-moving are significant aspects of minimising costs. Flexibility is vital: plans should be amenable to rapid change as decisions are made. The influences of special-interest groups and individual citizens require that plans be presented efficiently and in a manner that is easily understood. Simplified, visualised plans are instrumental in conveying both the content and the impact to those concerned.

Management and public services

In modern societies, decisions should be made quickly and on a reliable basis even though there may be many differing viewpoints to consider and a large amount of information to process. Today, the impact of decisions is ever greater, often because they involve conflicts between society and individuals, or between development and preservation. Information should therefore be readily available to decision-makers. As mentioned previously, 50% to 70% of the activities of all public administration involve some form of geographic information.

Overviews of administrative units and properties are crucial in the development of both virgin terrain and built-up areas, whether in developing or developed nations. In many countries, property registration is extensive: even in smaller states, two or three million properties may be involved. Moreover, real property is also an economic factor in taxation and as security for loans, so comprehensive overviews are essential to the well-ordered society. In the near future, computerised registers based on GIS technology are likely to be established in almost all countries.

Additionally, the printed media and TV increasingly use maps to convey information. As the media are shaping public opinion to an ever greater degree, it is important that it be supplied continuously with accurate and timely information.

Safety at sea

In many waters, the volume of shipping has increased considerably. Offshore oil rigs and other fixed facilities have been erected in customary shipping channels. As alterations along channels often affect navigation, there is an urgent need in shipping circles for up-to-date information. Today, a ship sailing in international waters may carry as many as 2,000 navigational charts, each of which may need updating.

Technology now permits sending information about navigational charts via public telecommunications networks, whilst electronic charts may be com-

Fig. 1.13 – A prerequisite for safe navigation at sea is rapid access to relevant navigational data. The risk of collision at sea is enhanced by the increasing speed of vessels and complexity of shipping lanes. Electronic sea chart systems function in the same way as GIS at sea and improve safety.

bined with positioning and radar displays on a ship's bridge. This means that ships can now sail more safely and at greater speeds in hazardous waters.

Transportation

In many countries, the greater part of transportation has shifted from rail and water to roads. At the same time, the use of private cars has multiplied. These developments have created traffic problems, which in turn cause loss of time and money. Large and sometimes hazardous goods are now transported by road. In most countries, the annual costs of traffic accidents have become extreme.

The transportation sector has always been a major consumer of maps and geographic data, so newer technologies may realise considerable savings.

The automobile industry is now investing heavily in the development of driver information systems. Several systems are now on the market, in a range of cars. In principle, all of them involve simple GIS functions with digital maps and supplementary information. In many urban areas where traffic is complex, increased driver information will ease congestion and improve safety.

Many countries have high-priority programmes for presenting digital road data for car navigation and transportation planning. This is because in most countries road transport costs are considerable and surveys have shown that the optimum choice of route and improved traffic flow can effect considerable savings. GIS may be used for both transportation planning and for the choice of optimum routes.

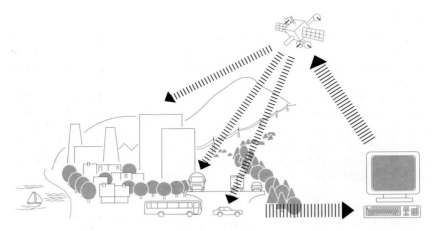

Fig. 1.14 – The transport sector has always made extensive use of geographic data. The introduction of new technology has opened for a new range of applications in the fields of transport planning and traffic information.

Military uses

Modern military products are extensively computerised. In many ways, military computer systems include GIS functions. Digital terrain data are used in flight simulators, aircraft navigation systems, weapons systems, command and control systems, and in flight mission planning systems. Many of these systems employ satellite images and Global Positioning System (GPS) satellite navigation data.

Military operations require a spectrum of information, much of it geo-referenced. Details may include terrain characteristics which are effective not only in locating radar and communications stations, but also the placement of fill for road building, the location of water sources, bridge load capacities and the location of strategic buildings, etc. All these data may be processed using GIS.

Military map agencies are increasingly using digital techniques. What is more, their input satellite data and satellite-aided positioning data are usually of a better quality than that in the civilian sector.

As a rule, military operations are relatively independent of civilian activities. However, the increasingly demanding military specifications pertaining to equipment and data access will undoubtedly benefit civilian GIS in due course. And trends in modern warfare indicate that military GIS activities can be expected to increase.

Conclusions

The age of information is upon us and those who fail to keep pace will soon be left behind. Hopefully, both producers and users of GIS will soon be aware of this trend, thus solidly establishing GIS amongst the foremost information technologies.

Considering the socio-economic trends discussed above, GIS can increasingly become a tool to be mastered in order to identify problems and outline solutions. Its technology has enormous potential, but it can also be abused. Messages to be conveyed may be altered unintentionally through incompetence or intentionally through conscious misuse. In many ways, GIS requires more of its users than traditional maps and archives.

1.3   BENEFITS OF COMPUTERISING INFORMATION

Staggering amounts of information are involved in almost all aspects of modern societies. GIS offers the advantage that it can process quantities of data far beyond the capacities of manual systems. Data in GIS are stored in a uniform, structured manner, as opposed to manual systems in which data are usually stored in various archives and files, in various agencies and organisations, on file cards, on various maps or in long reports. Therefore, in GIS data may be retrieved and superimposed far more rapidly than data in manual systems. In addition, data are quickly compiled into documents using techniques that include automatic map making and direct report printouts. The potential gains from switching from manually prepared maps and ordinary files to computerised GIS are considerable, both for public and private users.

A joint Nordic research project has evaluated the gains realised by implementing GIS through studies of its use in 50 to 60 agencies and organisations in the USA, Canada, Italy and the five Nordic countries – Norway, Sweden, Denmark, Finland and Iceland. The study showed that considerable benefits may be achieved provided the strategy used to implement GIS is suitably chosen. The study also showed that benefits are often related to objectives and that the following benefit-to-cost ratios may be attained by introducing GIS:

Fig. 1.15 – There are considerable gains to be made by transferring to computerised information. However, unlike with a slot machine, the more you play, the more you win.

Objective level	Map production	Map production and internal use of data	Map production, internal use of data and shared use of data
Tasks	• storage • manipulation • maintenance • presentation	• map production • planning • facility maintenance • project management	• map production • project • planning • facility maintenance • coordination • general service • facility management • economic planning • service and information
Benefit to cost ratio:	1:1	2 : 1	4 : 1

Fig. 1.16 – The benefit/cost ratio of transferring to GIS depends on how the system and information are applied. The ratio will normally be 1:1 for map production only; 2:1 for other applications such as planning etc.; and as much as 4:1 for a multiple-user information system.

1. If computerised GIS is used only for automated production and maintenance of maps, the benefit-cost ratio is 1:1.
2. If the system is also used for other internal tasks such as work manipulation and planning, the benefit-cost ration may be 2:1.
3. The full benefit of the system is first realised when information is shared among various users. The benefit-cost ration may then be 4:1.

The benefit-to-cost ratios quoted above refer to municipal services. Studies have shown that corresponding ratios for nationwide uses are somewhat lower, up to 2:1 to 3:1. Nonetheless, it's obvious that investment in GIS is as productive as investments in other sectors.

These benefits are not automatic. They depend largely on the proper choice of an introduction strategy. Even with the most suited strategy, there are no exact indices of benefits: the ratios discussed above are averages of spread empirical observations. Some figures, however, are impressive, with benefit-cost ratios of up to 8 to 10 or more. Detroit Edison (a public utility company in the USA) cut the production time for a particular type of map overview from 75 hours to one-and-a-half minutes. The city of Burnaby in Canada supported costs of a million dollars more than benefits for the first three years, but after seven years achieved an annual benefit of 400,000 dollars.

Some benefit-cost ratios for various sectors and activities are:

		Benefit-to-cost ratio
National data:	Forestry	2.0:1
	Transportation	4.0:1
	Environmental data	1.8:1
	Documentation	2.0:1

Municipal data: Map production   1:1
               Agency uses   2:1
               Joint uses   4:1

A number of GIS projects have been terminated because benefits have not been evaluated. The Nordic KVANTIF study showed that, characteristically, GIS projects involve high financial risk, but that successful projects often result in benefits greater than anticipated.

Benefits are a function of many factors. In mathematical notation, they might be written thus:

GIS Benefit = f (objectives, strategy, structure)

Objectives consist of visions and quantifiable goals. Strategy includes focus on selected products and tasks, the depth of investment, updating frequency and geographic coverage. Structure consists of organisational routines for the exchange of common data.

Systematic planning and conduct set profitable GIS projects apart from the unprofitable. Projects based on estimated cost calculations are often more profitable than projects driven by pure technology. Profitable projects are user oriented, rather than production oriented. Profitable projects start by being so clearly and convincingly defined that they are funded outside the ordinary operating budget.

The measurable benefits of GIS are usually expressed as gains in efficiency in terms of time saved, but there are also many cases of direct increases in income and reductions in costs.

Measurable benefits may include:

- improved efficiency due to:
  - more work being performed by the same staff
  - the same work performed by a smaller staff
- reduced current costs, with imminent costs precluded, due to such improvements as:
  - better bases for financial management
  - less costly maintenance of facilities
  - joint uses of available data
  - more efficient transportation
  - less excavation hauling
- income increases due to:
  - better subscriber/customer registers
  - improved property registers
  - sales of new products and services

Experience so far indicates that when GIS makes some jobs superfluous, staff are not made redundant but instead put to tasks that create more value (than just collecting data).

Intangible benefits may also accrue. They cannot be expressed directly in monetary terms, but should always be included when benefits are evaluated.

Intangible benefits may include:
- improved public and private decision-making in:
  – administration
  – planning
  – operations
- improved information and service to the public
- increased safety at sea
- improved environment for future generations
- better presentation of plans and their effects
- generally more "streamlined" socio-economic machinery which promotes economic growth

The greatest long-range global benefits of GIS are probably in the sectors where decisions have an environmental impact. The environment and the natural relationships within it are complex and not yet fully understood. It is, however, widely known that environmental degradation is implicated in the causes of many "modern" illnesses, such as asthma and cancer, the annual costs of which are enormous. Throughout history, battles have been fought and wars waged to gain access to scarce natural resources.

Increases in safety at sea are associated mainly with the introduction of electronic navigation systems, such as the Electronic Chart Disk Navigation System (ECDIS). The benefits realised can best be appreciated by considering what the ECDIS might help prevent. Among the more serious environmental disasters are oil spills from tankers that run aground. Subsequent cleanups have cost from to $40 per kilogram of oil, for totals of up to two billion dollars.

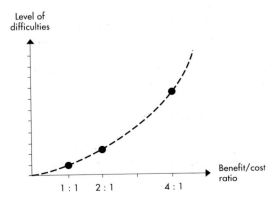

Fig. 1.17 – High benefit/cost ratio figures are not easy to achieve and the level of difficulty tends to increase proportionately.

If improved navigation using ECDIS can prevent one such disaster per year, the benefits are considerable, albeit difficult to assess directly.

Clearly, the assessment of benefits is seldom straightforward and the difficulties increase as the potential benefits become greater.

For further details on benefit-to-cost analyses and strategy choices, see Chapter 11.

1.4   USERS

Users may be classified in two groups. In the first group are operators who do not use the data themselves but process it for end users. In many cases, operator tasks may be compared to those of conventional cartographers or file clerks. GIS simplifies operator tasks and makes them more efficient.

The second group comprises end users and primary users who make decisions based on GIS products. The group includes:

- Operation and maintenance engineers; a typical decision may be whether to replace or repair a damaged water main.
- Regional planners; characteristic tasks involve presentations of plans to municipal authorities in a realistic, varied, visual manner.
- Building authority functionaries; representative jobs include processing building permit applications involving access roads, water supply, sewage, etc.
- Revenue functionaries typically dealing with tax assessment and taxpayer addresses.
- Road engineers, whose responsibilities include locating new roads in order to minimise cut-and- fill operations.
- Information officers; information produced may include complete packages to newly established firms with details on industrial areas, schools, transportation, etc.
- Local officials, who may require updated overviews on the effects of effluents on water quality at public beaches or the effects of zoning on school capacities.
- Fire brigades, for whom rapid, reliable information on the locations of fires and the presence of hazards such as explosives would be invaluable.
- Tourists who may seek such specific goals as geology study or wilderness experience.
- Forest manager planning harvest operations, computing volume of annual growth, estimating road cost and identifying sensitiv wildlife areas.

Fig. 1.18 – A possible future scenario with politicians grouped around a graphic "display-board" while discussing matters pertaining to geographic data.

- Bank officials, perhaps wishing to verify ownership of collateral.
- Oil tanker captains manoeuvring a ship in hazardous waters.
- Truck drivers seeking to minimise the problem of hauling an extra wide load between two points.

Such a list could be endless; only imagination sets a limit. Indeed, GIS may spawn information technologies applicable to completely new sectors, such as optimum warehousing or even brain mapping.

2　HISTORICAL DEVELOPMENT – GEOGRAPHICAL DATA AND GIS

2.1　EARLY DEVELOPMENTS

Geographical Information Systems evolved from centuries of map-making and compiling registers. The earliest known maps were drawn on parchment to show the gold mines at Coptes during the reign (1292 B.C.- 1225 B.C.) of Rameses II of Egypt. Perhaps earlier still are Babylonian cuneiform tablets that describe the world as it was then known.

Later, the Greeks acquired cartographic skills and compiled the first realistic maps. They began using a rectangular co-ordinate system for making maps around 300 B.C. About 100 years later, the Greek mathematician, astronomer and geographer Eratosthenes (ca. 276 – 194 B.C.) laid the foundations of scientific cartography. One of the earliest known maps of the world was

constructed by Claudius Ptolemaeus (c. A.D. 90 – 168) of Alexandria.

The Romans were more concerned with tabulations and registers. The terms cadastre (an official property register) and cadastral (of a map or survey that shows property or other boundaries) originate from the late Greek katà stíchon, which means "by line." But it was the Romans who first employed the concept to record properties, in the capitum registra, literally land register. In many countries, the term cadastre designates map and property registers. In France it also applies to the administrative agency in charge of such documents.

Throughout history, as societies organised, it became necessary to meet expenses. Some of the better known earlier examples include taxation levied by emperors and kings to

Fig. 2.1. – One of the oldest known maps in the world - a Babylonian cuneiform tablet dating from about 1400 B.C.

meet military expenses. These direct schemes comprised the foundations of today's complicated revenue systems involving the taxation, amongst other things, of income, property and goods. Both the ancient Egyptians and the Romans taxed property. Hence property registration was early systematised to assure tax revenues.

The earliest maps were drawn almost exclusively to facilitate commercial sea voyages. On them, coasts were meticulously detailed. Harbours were plumbed. Yet interiors remained unknown, apart from details of important trade and caravan routes.

The Arabs were the leading cartographers of the Middle Ages. European cartography degenerated as the Roman Empire fell. In the 15th Century, old skills were revived and Claudius Ptolemaeus's Geographia was translated into Latin to become the then existent view of the world. Although cartography was neglected, in many countries property registry thrived. The best known example is the Domesday Book, the record of the lands of England compiled in 1086 at the behest of William the Conqueror (1027 – 1087). The data included specifications of properties and their value, a count of inhabitants and live-stock, and their particulars, as well as incomes and taxes paid.

The travels and explorations of Marco Polo, Columbus, Vasco da Gama and others resulted in increased trade. In turn, maps were needed of previously unmapped seas and coasts. As the European countries and the newly dis-covered regions evolved to more organised societies, the needs for geographic information increased. Ordnance developments, such as the introduction of artillery, made maps important in military operations, and military agencies became the leading mapmakers. In many countries, the military mapmakers

Fig. 2.2 – This map, dating from 1570, is one of the oldest known maps of the Northern European seas. The cartographer was obviously more familiar with the southernmost than with the northernmost tracts.

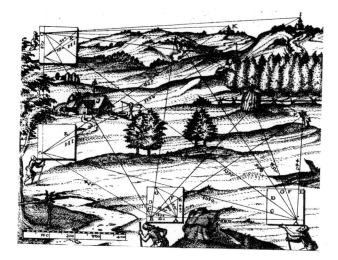

Fig. 2.3 – In the 18th century, vast land areas were surveyed using relatively exact methods. This task was often carried out by military mapmakers.

became responsible for all maps, civilian and military, both topographic land maps and navigational charts. Vestiges of this trend still remain: in many countries, map agencies, particularly nautical chart agencies, seem characteristically military. For instance, the official mapmaking agency of Great Britain and Ireland is the Ordnance Survey.

Until the 19th century, geographic information was used mostly in trade and exploration on land and sea, and for tax collection and military operations. New needs arose in step with evolving infrastructures, such as roads, railways, telegraph and telephone lines, gas and water supplies. Planning these facilities required information about the terrain beyond that commonly available. The accurate location of towns and cities, lakes and rivers, mountains and valleys became increasingly important. Detailed topographic information was needed to lay out railway and road grades and curve radii. Then, as now, foundations were a major challenge. So maps showing the type of soil, the quality, location and properties of bedrock were required. As planning advanced, specialised maps became more common. The first geological map of Paris was compiled in 1811. In 1838, the Irish Railway compiled a series of maps which may be regarded as the first manual geographic information system.

Development became increasingly dependent on socio-economic factors. The rights of property owners entered the picture because the construction of roads and railways often necessitated the expropriation of private lands. So new applications arose for property registers and maps, as builders needed to compile overviews of affected properties in order that their owners might be justly compensated.

As cities grew larger and more complex, accurate urban planning became a necessity. Many countries began compiling statistical information relating to urban planning in the early 19th century. By 1825, the British Ministry of Statistics had amassed extensive population statistics.

On the Continent, traditional village property ownership became a hindrance to effective farming. Many properties had become fragmented over the years owing to inheritance settlements. In some cases, a single property might comprise several hundred disperse parcels. Sometimes the ownership of, or rights to, a parcel were divided: one owner could have timber rights, another grazing rights, a third hunting rights, and so on. Therefore, property mapping of the late 19th century aimed to wrest order from chaos. With reference to available land registers, the various parcels were assembled into properties that were easier to work. Borders were consolidated, clarifying ownership and allowing property taxation.

Aerial photography accelerated the progress of map-making. The first aerial photo was used for mapmaking, and the first map-making instrument devised, in 1909. Photogrammetry, the technique of making measurements from photographs, developed rapidly in the 1920s and 1930s, and the two world wars also hastened developments. After the second world war, photogrammetry became widely used in mapmaking, mostly for maps in scales from 1:500 to 1:50,000. Aerial photographs themselves became important sources of quantitative information in evaluating such features as vegetation and geology.

2.2 THE FIRST AUTOMATIC PROCESSING OF GEOGRAPHIC INFORMATION

Although Baise Pascal is credited with devising the first true computing machine in 1642, large amounts of data were first processed automatically in 1890, when a tabulating device conceived by H. Hollerith was used in compiling the U.S. census. In Hollerith's first apparatus, census data were punched on cards which were then read electrically to compile data in separate registers. In the first half of the 20th century, Hollerith's various mechanisms were developed further. Data processing using punch cards became an industry.

During the second world war, data processing again advanced, primarily to meet the military need for predicting ballistic trajectories. One of the most famous computers developed for that purpose was EINAC, an acronym for Electronic Numerical Integrator and Calculator. After the war, computer development continued. By 1953 IBM had launched the model 650, which became the "Model T of the Computer Age" by virtue of being the first electronic computer not to be hand-made. More than 100 were produced – in those days an amazing quantity. In today's computer terms, EINAC, Whirlwind, the IBM 650 and other early electronic computers are referred to as "first generation." All first-generation computers suffered from a common drawback: they used vacuum tubes/valves which, like light bulbs, gave off heat and had limited lifetimes. That alone limited their applications. One 25,000-valve computer of the period was continuously manned by a staff of ten, of which two were technicians assigned to continuous replacement of burnt-out valves. Nonetheless, computerisation was established as the technology for processing large amounts of data. By 1952, all U.S. governmental statistical data was processed by electronic computers.

By the late 1950s and early 1960s, second-generation computers using transistors became available, outperforming their vacuum-valve predecessors. Suddenly, computers became affordable in disciplines other than those of major governmental agencies. Meteorologists, geologists and other geophysicists began using electronic map-making devices. Initially, the quality was poor, not least because automatic drawing machines had yet to be developed.

As the uses of second generation computers spread, theoretical models were evolved to use statistical data. Then, as now, public and private decision-making was often based on analyses of various classes of geographic data. These included demographic trends, cost-of-living variations, the distribution of natural resources, wealth and social benefits, and the demography of employment. In the early 1960s, the potential of electronic computers was recognised in Canada and the USA. In 1962, the Canadian Geographic Information System (CGIS) began operations and was thus world's first true GIS. Two years later, in the USA, a similar system, MIDAS, began processing data on natural resources.

The need for reliable geographic data multiplies with the expansion of road, rail, telecommunications and sewage networks, airports, electricity and water supplies, and other essential services vital to the infrastructure of urban areas. Terrain information on maps is now a vital planning tool, from the first conceptual stage to the final, legally binding plan. Burgeoning road networks have mandated extensive analyses of transport patterns. Since the mid-1950s, computers have been used in the USA to simulate traffic flows in relation to population distribution.

Often, conflicts of interest arise between developers and conservationists, or between municipal and regional planners and individuals. Most countries have laws intended to resolve such conflicts – planning and building statutes, environmental and historical preservation laws, and the like. Enforcement of these laws requires an improved overall view of property ownership. Many countries began computerising property registers in the mid 1970s. For instance, in Norway a project to compile a national computerised property register began in 1976. Designated by its purpose – Grunneiendommer, adresser og bygninger (GAB) (properties, addresses and buildings) – the register was fully operational by the late 1980s. Systems such as GAB may be viewed as the initial systems of Land Information Systems (LIS).

2.3 MICROPROCESSORS AND RECENT DEVELOPMENTS

In the 1960s and early 1970s, integrated circuits were developed and computer programs refined. The result: third-generation computers which brought computerisation to virtually all professional disciplines, especially those processing large amounts of data.

But the major breakthrough came in 1971-72 with the development of the microprocessor. In 1974, a microprocessor was used to build the first fourth-generation desktop computer. Seven years later, the first microprocessor-based desktop computer was launched as a "Personal Computer." By the mid-1980s, the computer field was divided into three categories according to size of computer: main frames, the descendants of the original large computers, intended for major data processing and computational tasks; Personal Computers (PCs), the increasingly ubiquitous desktop computers; and mini-computers, which were smaller than main frames but larger than desktop PCs. By the early 1990s, main frames had become physically smaller and computationally more capable. That trend was reflected strongly in PCs, which by 1990 were outperforming mini-computers built only a few years earlier. This development signalled the demise of the mini-computer, an event that was further hastened by the introduction of PC networks, in which processing and storage capacity may be shared and distributed. The overall trend is best illustrated in terms of the costs of computing: a computer's processing and storage capabilities which, in 1960, cost $100,000, could be purchased for $10 in 1984 and $0.50 in 1991. In

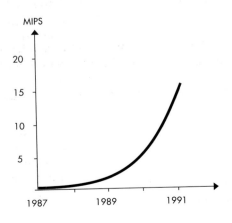

Fig. 2.4 – The figur illustrates how machin capasity in expression of MIPS (million instruction per second) have developed per US $10.000 from 1987 to 1991. One can see that computer's cost efficiency increase by a factor of 2 – 4 per annum.

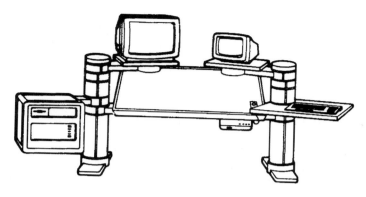

Fig. 2.5 – A work
station, with graphics
display, digitiser,
secondary storage
and data processor.
(SysScan)

other words, cost efficiency increased by a factor of ten every two to three
years.

In the 1970s and 1980s, various systems were evolved to replace manual
cartographic computations. Workable production systems became available in
the late 1970s and system development continued through the 1980s. Nonetheless, by the early 1990s, elegant approaches to some cartographic tasks have
yet to be found, and computerised cartographic research and development
remains a continual challenge.

The spread of PCs spurred user-friendly operations and programs capable
of processing in ways previously not possible, for example by considering the
logical connections in geographic data.
The increases in microprocessor computing capacity also made the processing of digital and satellite images and other types of raster images commercially available in the mid 1980s.

Software systems have developed apace. Relational databases systems, such
as dBase and Oracle which first appeared in the late 1980s, are particularly
useful in processing geographic data. Commercially available relational databases are now routinely used in GIS systems.

In the late 1980s, computing capability became widely accessible as microprocessors were used for multitude of devices, from household appliances and
automobiles to an extensive range of specialised instruments, including those
used in GIS. For GIS users, microprocessors have improved such devices as:

- surveying instruments
- GPS (Global Positioning System)
- digitising table
- scanners
- environmental monitoring satellites
  and data presentation systems, including:

- graphic displays
- electrostatic plotters
- laser printers

The various devices and sub-systems employed in GIS are discussed in Chapter 5.

3.   FROM THE REAL WORLD TO GIS

3.1   THE REAL WORLD

In many ways GIS presents a simplified view of the real world. Yet the processes involved are seldom straightforward because realities are irregular and constantly changing, so perception of the real world depends on the observer. For instance, a surveyor might see a road as two edges to be surveyed whilst the road works authority might regard it as an asphalt surface to be maintained, and the driver as a carriageway. Moreover, the real world may be described in terms of countless phenomena, from basic sub-atomic particles up to the dimensions of oceans and continents. The complexity and enormity of the real word, combined with a whole spectrum of interpretations of it, imply that the designs of GIS systems may vary according to the capabilities and preferences of their creators. This human factor can introduce an element of constraint, as data compiled for a particular application may be less useful elsewhere.

The systematic structuring of the data determines its ultimate utility and consequently the success of the relevant GIS application. This aspect is also characteristic of the data available in traditional maps and registers.

The real world can be described only in terms of models which delineate the *trace the outline of* concepts and procedures needed to translate real world observations into data that are meaningful in GIS. The process of interpreting reality by using a real world model and a data model is called data modelling. The principles involved are illustrated in Fig. 3.1.

3.2   REAL WORLD MODEL

The arrangement of the real world model determines which data needs to be acquired. The basic carrier of information is the entity, which is defined as a real world phenomenon that is not divisible into phenomena of the same kind.

An entity consists of: • type classification
                        • attributes
                        • relationships

Entity types

The concept of entity types assumes that uniform phenomena can be classified. During the classifaction process, each entity type must be uniquely defined to preclude ambiguity. For instance, house must be defined in such a way that detached house at No. 10 Church Road is classified under house and not under industrial building.

possibility of interpreting an expression in more than 1 way

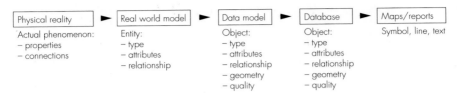

Fig. 3.1 – In order to bring the real world into GIS, one has to make use of simplified models of the real world. Uniform phenomena can be classified and described in the "Real World Model". The Real World Model is converted into a "Data Model" by applying elements of geometry and quality. The data model is transferred to a "Data base" which can handle digital data, from which, in turn, the data can be presented on maps and in reports. (Cederholm, Petterson, 1989)

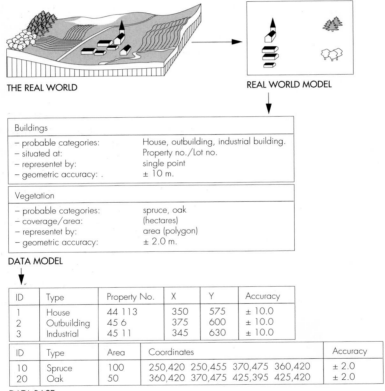

THE REAL WORLD

REAL WORLD MODEL

Buildings	
– probable categories:	House, outbuilding, industrial building.
– situated at:	Property no./Lot no.
– representet by:	single point
– geometric accuracy: .	± 10 m.

Vegetation	
– probable categories:	spruce, oak
– coverage/area:	(hectares)
– representet by:	area (polygon)
– geometric accuracy:	± 2.0 m.

DATA MODEL

ID	Type	Property No.	X	Y	Accuracy
1	House	44 113	350	575	± 10.0
2	Outbuilding	45 6	375	600	± 10.0
3	Industrial	45 11	345	630	± 10.0

ID	Type	Area	Coordinates				Accuracy
10	Spruce	100	250,420	250,455	370,475	360,420	± 2.0
20	Oak	50	360,420	370,475	425,395	425,420	± 2.0

DATA BASE

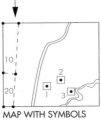

MAP WITH SYMBOLS

Fig. 3.2 – Modelling process. The transformation of the real world into GIS products is achieved by means of simplification and models.

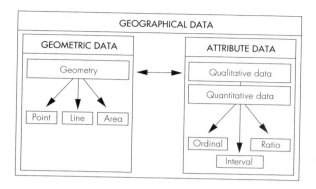

Fig. 3.3 – Geographical data can be divided into geometric data and attribute data. Attribute data can in turn be sub-divided into qualitative data and quantitative data.

Some organisations may need to classify entity types into categories as well as entities according to type. For instance, national highways, country roads, urban roads and private roads might come under the roadways category, or all entities within a specific geographic area might belong to a unique category.

Entity attributes

Each entity type may incorporate one or more attributes that describe the fundamental characteristics of the phenomena involved. For instance, entities classified as buildings may have a material attribute, with legitimate entries frame and masonry and a number of storeys attribute with legitimate values of 1 to 10, etc.

In principle an entity may have any number of attributes. For instance, a lake may be described in terms of its name, depth, water quality or fish population as well as its chemical composition, biological activity, water colour, algae density, potability or ownership.

Entity types may also describe qualitative data and attributes quantitative data. In principle, quantitative data may be ranked in three levels of accuracy. The most accurate are proportional, such as length and area which are measured with respect to an origin or starting point. Interval data, such as age and

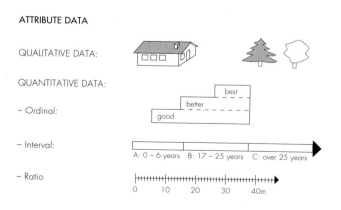

Fig. 3.4 – Attribute data consists of qualitative or quantitative data. Qualitative data specifies the type of object, whilst quantitative can be categorised into proportional data, i.e. data measured in relation to a starting-point "0", interval data which is data arranged into classes and ordinal data, which specifies quality by use of text.

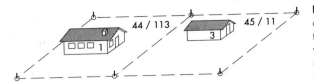

Fig. 3.5 – Relations. The computer cannot "see" the real world. It is therefore necessary to specify the varioyus relations between entities such as: belong to, comprise, are located in/on or border on.

income, comprise groups and are thus less accurate. The least accurate are ordinal data of rank, like "good", " better" and "best".

Entity relations

Relations often exist between entities. Typically, these include:

Relation:	Typical example(s):
– pertains/belongs	– a depth figure pertains to a specific shoal, or a pipe belongs to a larger network of contiguous pipes.
– comprises	– a country or a state comprises counties, which in turn comprise townships.
– located in/on	– a particular building is located on a specific property.
– borders on	– two properties have a common border.

Whilst such relations are intuitively obvious on ordinary maps, computers have no intuition. The computer processing of relations therefore requires either further descriptive information or instructions on how it may be compiled. As an aside, this aspect of map reading – the human facility of being able to see what a computer needs to be told to "see" – highlights pivotal differences between human and computer processing. As a sign in one international computer research laboratory states: "Computers: Fast, Accurate, Stupid. Humans: Slow, Slovenly, Smart."

Some relations may be unwieldy. One cause may be complexity, as in networks where the states of switches or valves, open or closed, determine which parts of the network may viewed as comprising a logical entity. In such situations, one may differentiate between actual and potential relations.

3.3   DATA MODEL

A real world model facilitates the study of a selected area of application by reducing the number of complexities considered. Those outside the selected area are considered insignificant. However, if a real world model is to be of any use, it must be realised in a database. The data model makes that possible.

Unlike humans, computers cannot learn the essentials of manholes, property lines, lakes or other types of objects. What they can do is to manipulate geometric objects such as points, lines and areas, which are used in data models.

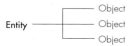

Fig. 3.6 – A single entity can be described by several objects, i.e. there are many relationships between entities.

The carriers of information in data models are known as objects. These correspond to entities in real world models and are therefore regarded as database descriptions of real-world phenomena.

Objects are characterised by:
- type
- attributes
- relations
- geometry
- quality

Real-world models and entities cannot be realised directly in databases, partly because a single entity may comprise several objects. For instance, the entity Church Road may be represented as a compilation of all the roadway sections between intersections, with each of the sections carring object information.

Multiple representations produced by such divisions may promote the efficient use of GIS data.

Therefore, information-carrying units and their magnitudes must be selected before the information is entered in a database. For instance, the criteria for dividing a roadway in sections must be selected before the roadway can be described.

Objects

Objects in a GIS data model are basically described in terms of identity type, geometric elements, attributes, relations and qualities.

Identities, which may be designated by numbers, are unique: no two objects have the same identity. Type codes are based on object classifications which can usually be transferred from entity classifications. An object may classify under one type code only.

Data models may be designed to include:
- physical objects, such as roads, water mains and properties,
- classified objects, such as types of vegetation, climatic zones or age groups,
- events, such as accidents or water leaks,
- continously changing objects, such as temperature limits,
- artificial objects, such as elevation contours and population density,
- artificial objects for a selected representation and database (raster).

3.3.1   GRAPHICAL REPRESENTATION OF OBJECTS

Graphical information on objects may be entered in terms of:
- points (no dimensions)
- lines (one dimension)
- areas (two dimensions)

Points

A point is the simplest graphical representation of an object. Points have no dimensions but may be indicated on maps or displayed on screens by using symbols.

The corner of a property boundary is a typical point, as is the representative co-ordinate of a building. That said, it is of course the scale of viewing which determines whether an object is defined as a point or an area. In a large scale representation a building may be shown as an area whereas it may only a point (symbol) if the scale is reduced.

Lines

Lines connect at least two points and are used to represent objects which may be defined in one dimension. Property boundaries are typical lines. So are electric power lines and telecommunications cables. Roads and rivers, on the other hand, may be either lines or areas, depending on the scale.

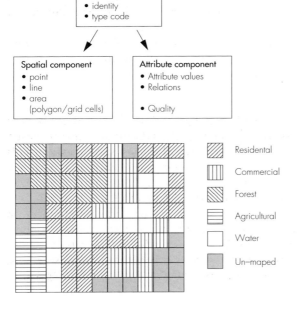

Fig. 3.7 – In a data model, objects are categorised as object classifications, geometric elements (point, line, area), attributes, relations between the entities and quality definitions of these descriptive elements.

Fig. 3.8. – This figure illustrates land-use in a town, in the form of a raster map. The land-use is registered in a raster system with cells of equal size. Each category of land use is given its own symbol on the map.

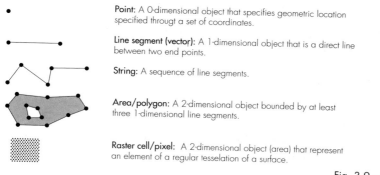

Point: A 0-dimensional object that specifies geometric location specified througt a set of coordinates.

Line segment (vector): A 1-dimensional object that is a direct line between two end points.

String: A sequence of line segments.

Area/polygon: A 2-dimensional object bounded by at least three 1-dimensional line segments.

Raster cell/pixel: A 2-dimensional object (area) that represent an element of a regular tesselation of a surface.

Fig. 3.9 – Point, line, area.

Areas

Areas are used to represent objects defined in two dimensions. A lake, an area of woodland or a township may typically be represented by an area. Again, physical size in relation to the scale determines whether an object is represented by an area or by a point.

An area is delineated by at least three connecting lines, each of which comprise points. In databases, areas are represented by polygons, i.e. plane figures enclosed by at least three straight lines intersecting at a like number of points. Therefore, the term polygon is often used instead of area.

Reality is often described by dividing it into regular squares or rectangles so that all objects are described in terms of areas. This entire data structure is called a grid. Population density is well suited to grid representation. Each square or rectangle is known as a cell and represents a uniform density or value. The result is a generalisation of reality, depending on cell size.

All cells of a grid in a data model or a database are of uniform size and shape, but have no physical limits in the form of geometric lines.

Objects in three dimensions

All physical phenomena have locations in space; thus, the world can be said to be three-dimensional. Consequently, a complete data model must be based on three dimensions: the ground plane (Northing/Easting) and elevation. This applies not only to terrain surfaces but also to towers, wells, buildings, borders, addresses, accidents and all manner of data. A complete data model must therefore manipulate geo-referenced data in three dimensions.

In many cases elevation may be regarded as an attribute of an object, in others as part of a graphical representation. For instance, contour lines connect points of like elevation, so elevation may be regarded as an attribute of a contour line. For a line representing a road, along which the elevation varies, elevations are more suitably regarded as part of the graphical description in three dimensions.

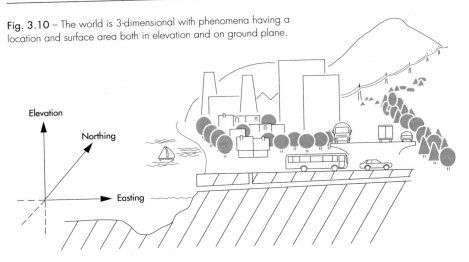

Fig. 3.10 – The world is 3-dimensional with phenomena having a location and surface area both in elevation and on ground plane.

In short, applications determine representations. However, practical limitations also influence choices. For instance, specification of all objects in three dimensions can easily multiply the amount of data collected beyond that which is needed. It may also influence the techniques used to collect data. The collection of photogrammetric data provides "free" elevation data in addition to the Northing and Easting data of the ground plane. When existing maps are digitised, however, elevation data must be entered manually (and sometimes inaccurately as exact elevations may not be available for all objects.)

In general, GIS permits the inclusion of the third dimension of height or elevation to a greater degree than does conventional cartography.

Drawbacks of graphical representations

Regarding the real world as comprising geometric figures (points, lines, areas) means viewing objects as discrete data model representations. That is to say, all objects have clearly defined physical limits. These limitations are most obvious on maps, where lines imply sharp demarcations with no smooth, continuous transitions.

A discrete data model does not always suit reality. For instance, exact elevations are seldom observed on the ground because elevation generally varies continuously. Nevertheless, it is normal to use discrete contour lines for map elevations.

Difficulties also arise in depicting phenomena that lack clear physical demarcation, such as soil types, population densities or prevailing temperatures. These problems can be solved by defining boundaries and establishing variable or fixed areas in which attribute values are presumed constant.

Object	Attributes				
no.	A	B	C		X
1	////	////			
2	////	////	////		////
3	////				
–					
–					
–					
n	////		////		

Fig. 3.11 – Attributes can be tabulated with each individual object represented by a line and each attribute by a column.

Cell	Attributes				
no.	A	B	C		X
01	////	////			
02	////	////	////		////
03	////				
11					
12					
–					
nn	////		////		

Fig. 3.12 – In principle, the difference between vector data and raster data is not that great. Raster data could well be arranged in tabular form with each cell number representing a line and each attribute (layer of raster values) a column.

3.3.2   OBJECT ATTRIBUTES

Object attributes are the same as the entity attributes of the real-world model. Attributes describe an object's features and may thus be regarded as a computer's "knowledge" of the object. In practice, object attributes are stored in tables, with objects on lines and attributes in columns.

Attribute values connected to grid data can theoretically be presented in the same way. Each grid cell corresponds to an object.

3.3.3   OBJECT RELATIONS

Object relations are the same as entity relations in the real-world model. One differentiates between:

1. Relations which may be calculated from
  • the coordinates of an object, for instance which lines intersect or which areas overlap, and
  • object structure (relation), such as the beginning and end points of a line, the lines that form a polygon or the locations of polygons on either side of a line.
2. Relations which must be entered as attributes, such as the division of a county into townships or the levels of crossing roads that do not intersect.

3.3.4   OBJECT QUALITY

The true value of any description of reality depends on the quality of all the data it contains – graphical, attributes and relations. Graphical data accurate to

± 0.1 meter obviously describes reality more faithfully than data accurate to ± 100 meters. Similarly, recently updated data is preferable to five-year-old data (which brings in temporal factors).

In the initial data modelling stage, the kind of considerations given to data quality should include:

- graphical accuracy (such as ± 1.0 m accuracy)
- updating (when and how data should be updated)
- resolution/detailing – for instance, whether roads should be represented by lines or by both road edges
- extent of geographical coverage, attributes included, etc.
- logical consistency between geometry and attributes
- representation: discrete vs. continuous
- relevance – for example, where input may be surrogate for original data that is unobtainable.

Information on the quality of data is important to users of the database. Requirements for data quality are discussed in greater detail in Chapter 7.

3.4   FROM DATABASE TO GIS TO MAP

Once a data model is specified, the task of realising it in a computer is technical and the task of entering data simple and straightforward, albeit time consuming.

A database need seldom be made to suit a data model as many databases compatable with GIS applications are now on the market. The problem at hand is more one of selecting a suitable database with regard to:

- acquisition and control
- structure
- storage
- updating and changing
- managing and exporting/importing
- processing
- retrieval and presentation
- analyses and combinations

Needless to say, a well prepared data model is vital in determining the ultimate success of the GIS application involved.

Users view reality using GIS products in the form of maps with symbols, tables and text reports. The dissemination of information via maps is discussed further in Chapter 10.

4. DIGITISING THE REAL WORLD

4.1 GENERAL

As discussed in Chapter 3, GIS depicts the real world through models involving geometry, attributes, relations and data quality. In this Chapter, the realisation of models will be described, with the emphasis on geometric spatial information, attributes and relations. Data quality will be discussed further in Chapter 7.

Spatial information is presented in two ways: as vector data in the form of points, lines and areas (polygons) or as grid data in the form of uniform, systematically organised cells. Geometric presentations are commonly called digital maps. Strictly speaking, a "digital map" would be peculiar because it would comprise only numbers (digits). By their very nature, maps are analogue, regardless of whether they are drawn by hand or machine, or whether they appear on paper or are displayed on a screen. Technically speaking, GIS does not produce "digital maps:" it produces analogue maps from digital map data. Nonetheless, the term digital map is now so widely used that the distinction is well understood.

The manner in which digital map data are stored in a record is determined by a format, a set of instructions specifying how data are arranged in fields. The latter are groups of characters or words which, in turn, are treated as units of data. The format stipulates how the computer shall read data into the fields; the total number of fields specified, the number of characters permissible in each field, the number of spaces between fields, which fields are numeric and which are text, and so on. An example is shown in Fig. 4.1.

The information content of the data is designated not in the format but ancillary to it, for example in a heading. Typical specifications for information content might include field assignments, such as the point number in the first field, the numeric code in the second, Easting in the third, Northing in the fourth and elevation in the fifth. Needless to say, the meanings of the numeric codes used must also be given.

	Field 1	Field 2	Field 3
Record 1	110	Oslo	10.61
Record 2	115	Stockholm	9.15
Record 3	116	London	18.33
—			
—			
Record n			

Fig. 4.1 – In a single data file, data are stored in records and each record is divided into fields where the different values are stored. The format determines how the data are arranged in fields, according to the number of characters, whether they are text or numeric etc.

4.2   VECTOR DATA MODEL
4.2.1   GENERAL

In principle, every point on a map and every point in the terrain it represents may be uniquely located using two or three numbers in a coordinate system, such as in the Northing, Easting and elevation Cartesian coordinate system. On maps, coordinate systems are commonly displayed in grids with location numbers along the map edges. On the ground, coordinate systems are imaginary yet marked out by survey control stations. Data usually may be transformed from one coordinate system to another. See Chapter 4.9 for further details on coordinate systems.

With few exceptions, digital representations of spatial information in a vector model are based on individual points and their coordinates. The exceptions include cases where lines or parts of lines (for instance, those representing roads or property boundaries) may be described by mathematical functions, such as those for circles or parabolas. In these cases, GIS data include equation parameters – for example, the radii of the circles used to describe parts of lines.

Together with the coordinate data, instructions are entered on which points in a line are unconnected, and which are connected. These instructions can subsequently be used to trigger "pen up" and "pen down" functions in drawing.

Coordinate systems are usually structured so that surveys along an axis register objects in a scale of 1:1; that is to say, one metre along the axis corresponds to one metre along the ground. In principle, measurements along an axis can be as accurate as one wishes, but the degree of precision will naturally influence the amount of work required to gather the data.

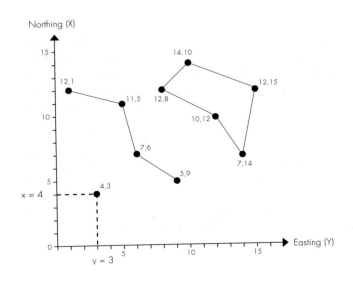

Fig. 4.2 – In the Vector Data Model, the objects are described as points, lines or areas (polygons). These three geometric phenomena may individually be described by a single point in a coordinate system and with connected lines (for lines and areas). Each object therefore comprises one or several sets of coordinates.

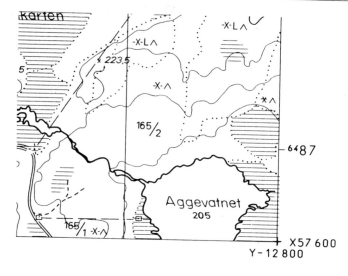

Fig. 4.3 – On maps, coordinate systems in an x- and y- direction, are generally displayed in grids, with location numbers along the map edges.

Mathematically, a vector is a straight line, having both magnitude and direction. Therefore, a straight line between two data coordinate points on a digital map is a vector. Hence the concept of vector data used in GIS and the designation of vector-based systems.

In the vector model, points, lines and areas (polygons) are the homogeneous (and discrete) units which carry information. As discussed above, these three types of object may be graphically represented using coordinate data. However, as we shall see, the objects may also carry attributes which can be digitised, and all digital information can be stored.

4.2.2 CODING DIGITAL DATA

Anyone familiar with maps knows that map data are traditionally coded. Roads, contour lines, property boundaries and other data indicated by lines are usually shown in lines of various widths and colours. Symbols designate the locations of churches, airports, and other buildings and facilities. In other words, coordinates and coding information identify all objects shown on a map.

Not surprisingly, then, the digital data used to produce maps is also coded, usually by the assignment of numeric codes used throughout the production process – from the initial data to computer manipulation and on to the drawing of the final map. Typical numerical coding is illustrated in Fig. 4.4.

Each numerical code series contains specific codes assigned to the objects in the group. For instance, the codes for boundaries may be as illustrated in Fig. 4.5.

Numerical code series	Object group
1 000	Survey control stations
2 000	Terrain formation
3 000	Hydrography
4 000	Boundaries
5 000	Built-up areas
6 000	Buildings and facilities
7 000	Communications
8 000	Technical facilities

Fig 4.4 – Digital map data often uses numerical coding, in the form of different numerical series, to identify object groups.

Numerical code	Object type
4 001	National border
4 002	County boundary
4 003	Township boundary
4 011	Property boundary
4 022	National park border
	etc.

Fig. 4.5 – By using a numerical coding system, codes can be assigned to all levels of detailed information on the objects.

Digital data for map production comprises sequences of integers, such as:

– 5314401112345678912340 6780–5314401112333 6788123306700

Use of the format permits the numerical sequence to be divided into groups and read:

– 5/314/4011/12345/6789/12340/6780
– 5/315/4011/12333/6788/12330/6700

The figures designate:

Figure	Designates
–5	– start of a continuous sequence of data (i.e., if there are several – coordinates, they shall be connected in a line: "pen down")
314	– serial number of data sequence (such as of a unique line)
4011	– property boundary (such as might produce a final line width of 0.3 mm)
12345	– first Easting coordinate
6789	– first Northing coordinate (pen moves to next coordinate set)
12340	– last Easting coordinate
6789	– last Northing coordinate
–5	– end of data sequence, start of next sequence ("pen up" – moved and set down for a sequence of new coordinates, and so on)

In thematic coding, which may be compared to the overlay separation of conventional map production, data are divided into single-topic groups, such as all property boundaries. Information on the types of symbols types, line widths, colours etc. may be appended to each thematic code, and various combinations of themes may be drawn.

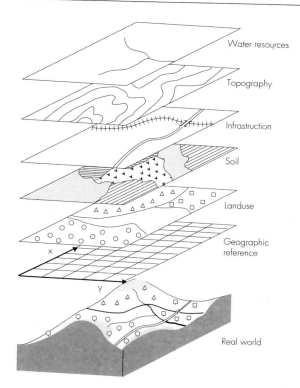

Water resources

Topography

Infrastruction

Soil

Landuse

Geographic
reference

Real world

Fig. 4.6 – Theme codes in the digital map data can be used to separate data into thematic layers.

Data may be jointly presented in this way only if all objects are registered, using a common coordinate system.

Coding digital data for GIS

Point objects may easily be realised in a database because a given number of attributes and coordinates is associated with each point; see Fig. 4.7.

Line and polygon objects are more difficult to realise in a database because of the variation in the number of points composing them. A line or a polygon may comprise two points or 2,000 – or more – points, depending on the extent of the line and the complexity of the area, which is delineated by a boundary line that begins and ends at the same point.

Object spatial information and object attributes are often stored in different databases in order to ease the manipulation of lines and areas, but in some systems they are stored together. As pivotal attributes are often available in existing computer memory files, dividing the databases conserves memory by precluding duplicate storage of the same data.

The separate storage of attribute and spatial information data requires that all objects in the attribute tables be associated with the corresponding spatial information. This association is achieved by inserting spatially stable and rele-

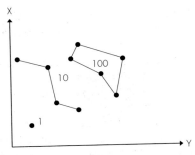

Fig. 4.7 – Each object is assigned attributes and coordinates. However, the number of coordinates for lines and polygons will vary considerably, depending on the length of the line and circumference of the polygon. This may make it inappropriate to store attributes and geometry together, and is one of the reasons why many systems store them separately.

Geometry	ID	Attributes		Coordinates
Point	1	A	B	4,3 (single point)
Line	10	C	D	2,1 11,5 · · 5,9 (string)
Area	100	C	E	4,10 12,14 · · · · 14,10 (closed polygon)

0-DIMENSIONAL OBJECT:

Point: A 0-dimensional object that specifies geometric location specified througt a set of coordinates.

Node:.A 0-dimensional object that is a topological junction and may specify geometric location.

1-DIMENSIONAL OBJECTS

Line segment (vector): A 1-dimensional object that is a direct line between two end points.

Link: A 1-dimensional object that is a direct connection between two nodes.

Directed link: A link between two nodes with one direction soecified.

String: A sequence of line segments.

Chain: A directed sequence of nonintersecting line segments with nodes at each end.

Arc: A locus of points that forms a curce that is defined by mathematical function. Also defined as string or chain.

Ring: A sequence of any line segments with closure.

2-DIMENSIONAL OBJECTS

Simple area/polygon: An area defined by an outer ring, but that may not have inner rings (holes).

Complex area/polygon: An area defined by an outer ring with optional inner rings defing holes.

Raster cell/pixel

Fig. 4.8 – Geometric objects. (adapted from Agnar Renolen))

vant attribute data or codes from the attribute table into the spatial information, or vice versa. In other words, identical objects have the same identities in both databases.

The identity codes used to label and connect spatial information and attribute table data are most often numeric, but may be alphanumeric. Typical identity codes include building numbers, property numbers, street numbers and addresses. If the data are ordered in a manuscript map, each object may be assigned a serial number used both in the spatial information and the attribute databases. Polygons for vegetation mapping can, for example, be numbered from 1 onwards, whilst pipes, manholes, etc. are usually numbered according to an administrative system.

Spatially defined objects without attributes need no identifiers, but they are required for all objects which are listed in attribute tables, and manipulated spatially.

Identifiers are normally entered together with the relevant data, but they may also be entered later, using an interactive man-machine process such as keying in identifiers for objects pointed out on the screen.

Some systems tie attributes to characteristic points in polygons. These points may be computed or identified interactively on screen, and codes may be entered manually for the relevant polygons.

Typical digital data for GIS are illustrated in Fig. 4.9.

I.D.	Thematic code	X-coordinates	Y-coordinates
11	30	23 999.80	10 008.55
	30	23 990.50	10 015.10
34	40	24 876.30	11 122.86
	40	24 890.10	11 150.30
–	–	–	–
	–	–	–
122	20	24 870.25	11 130.23

Code list	
20	Topography
30	Infrastructure
40	Soil

Fig. 4.9 – This figure shows a typical section of digital map data with relevant code list.

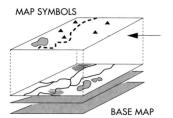

MAP SYMBOLS

BASE MAP

Look-up table	
Thematic code	Map symbols
20	triangle
30	dotted line
40	shaded area

Table		
ID	Thematic code	Coordinates
	20	
	30	
	40	

Fig. 4.10 – Drawing instructions are often designated in look-up tables. Thematic code values or attribute values are often input values in the tables, whereas output values can be symbol types, colours, line thicknesses etc.

The thematic codes are attributes stored together with the coordinates.

Plotting may be controlled by appending drawing instructions to the thematic code or to the individual identifications, or to other object attribute values. In a finished map, tabular data appear on a "foreground map" against the "background" of a base map derived from the remaining map data. Look-up tables are usually used to translate tabular data to map symbols. Map presentations are discussed further in Chapter 10.

4.2.3   STORING VECTOR DATA

Users of conventional maps know the frustrations of extracting information from maps produced by various agencies using differing map sheet series, varying scales and coordinate systems and, frequently, different symbols for the same themes. Moreover, the cartographic version of "Murphy's Law" seems to dictate that the information of interest is all too often located in the corners where four adjoining map sheets meet.

Database storage of cartographic data can overcome these problems because it involves standardisation of data through common reference systems and uniform formats. Cartographic data from various sources can, with few limitations, be combined. The results are then independent of map sheet series and scales.

Standardised storage makes the presentation of data compiled from dissimilar sources much easier. For instance, uniform storage formats permit the combination of telecommunications administration network data with property survey data, or of geological information from 1:50,000 scale maps with vegetation data from 1:20,000 scale maps.

Digital map data are stored in databases, the computerised equivalent of conventional file drawers and cabinets. Although data entries in a database may be updated far more rapidly than data printed on map sheets on file, the information available on map sheets is usually more rapidly available than that contained in a database. This is because a single map sheet contains an enormous amount of information, usually equivalent to 100,000 or more sets of coordinates. A sequential computer search of 100,000 items in a database is lethargic even for the most powerful of computers in comparison with a quick visual scan of a map sheet. Therefore, "smart" programs known as Database Management Systems (DBMS) have been compiled to maintain, access and manipulate databases. The various DBMS differ primarily in the ways in which the data are organised. Their selection and use are vital in GIS applications because they determine the speed and flexibility with which data may be accessed. Further details on cartographic databases are given in Chapter 8.

In practice, vector data are stored either by using the "spaghetti" (unlinked) model or the topology model.

Spaghetti model

As illustrated earlier in this chapter, digital map data comprise lines of contiguous numerals pertaining to spatially referenced points. "Spaghetti" data are a collection of points and line segments with no real connection. What appears as a long, continuous line on the map or in the terrain, may consist of several line segments are to be found odd places in the data file. There are no specific points that designate where lines might cross, nor are there any details of logical relationships between objects. Polygons are represented by their circumscribing boundaries, as a string of coordinates, so that common boundaries between adjacent polygons are registered twice (often with slightly differing coordinates). The lines of data are unlinked and together are a confusion of crossings.

Unlinked, or "spaghetti" data usually include data derived either from the manual digitising of maps or from digital photogrammetric registration. Consequently, spaghetti data are often viewed as raw digital data. These data are amenable to graphic presentation – the delineation of borders, for example – even though they may not form completely closed polygons. Otherwise their usefulness in GIS applications is severely limited.

One drawback is that both data storage and data searches are sequential . Hence search times are often unduly long for such routine operations as finding commonality between two polygons, determining line intersection points or identifying points within a given geographical area. Other operations vital in GIS, such as overlaying and network analysis, are intractable. Furthermore, unlinked data requires an inordinate amount of storage memory because all polygons are stored as independent coordinate sequences, which

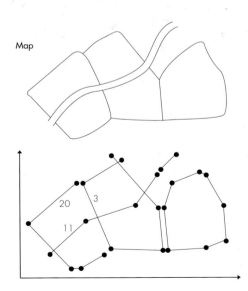

Map

Fig. 4.11 – "Spaghetti" data is the term which is often used to describe digital map data with crossing lines, loose ends, double digitalisation of common boundaries between adjacent polygons etc. These data lie in a pile, just like spaghetti. Several line segments (which) are to be found odd places in the data file..

Table

Line nr.	Coordinates
1	...
2	...
3	X, Y, Z
.	...
.	...
.	...
11	X, Y, Z
.	...
.	...
20	X, Y, Z

Map scale	Map sheet	MBytes
1 : 250 000	50 x 60 cm	25 – 50
1 : 50 000	50 x 60 cm	15 – 25
1 : 5 000	48 x 64 cm	2,5 – 10
1 : 1 000	60 x 80 cm	1 – 3

Fig. 4.12 – Digital map data requires an inordinate amount of storage memory. Normally elevation data comprises 25-90% of the total data volume.
This example is taken from a Norwegian map series.

means that all lines common to two neighbouring polygons are stored twice. The typical memory required for unlinked data is illustrated in Fig. 4.10.

Data compression

The amount of memory needed can be reduced by using data compression techniques. Most of these involve the elimination of repetitive characters – for example, the first character of all coordinates along a particular axis. The repetitive character needs to be entered only once; subsequently it may be added to each set of coordinates. Savings in characters stored are illustrated in Fig. 4.13.

There are other memory-saving devices, too. Contours and other lines can be replaced with relatively simple mathematical functions, such as straight lines, parabolas and polynomials. A spline function comprises segments of polynomials joined smoothly at a finite number of points so as to approximate a line. As shown in Fig. 4.14, a spline function involves several such polynomials to build a complex shape. A spline function representing nautical chart data has been reported to reduce data volume by 95%. (Dahlen & Holm, 1991)

Additionally, points of little or no value in describing a line may be eliminated by moving a corridor along a line and deleting points outside it. The width of the corridor determines which points are deleted.

The amount of memory required to store a given amount of data often depends on the format in which data are entered. Some formats contain more administrative routines than others. Some have vacant space. This means that the gross volumes stored are frequently related to format.

Topology model

Topology is the branch of mathematics that deals with geometric properties which remain invariant under certain transformations such as stretching

Original data		Compacted data	
Northing	Easting	10,000	70,000
10,234	70,565	234	565
10,245	70,599	245	599
10,167	70,423	167	423
–	–	–	–
–	–	–	–
etc.	etc.	etc.	etc.

Fig. 4.13 – Simple data compression. The volume of data to be stored can be reduced to a single entry giving the value common to all coordinate values.

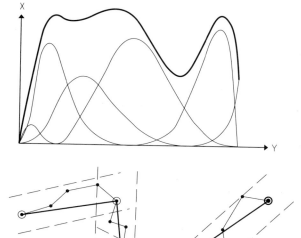

X

Y

Fig. 4.14 – A complicated line course can be replaced by a few simple polynomials - spline. This reduces considerably the volume of data emanating from line courses.

Fig. 4.15 – Reduce lines by use of a corridor. The number of points needed to describe a line may be reduced by moving forward a corridor of a given width until it touches the digitalised line. All the points on the line in the corridor, apart from the first and last, are thereby deleted. This process is repeated until the whole line has been trimmed.

or bending. The topology model is one in which the connections and relationships between objects are described independently of their coordinates: their topology remains fixed as geometry is stretched and bent. Hence the topology model overcomes the major weakness of the spaghetti model, which lacks the relationships requisite to many GIS manipulations and presentations.

The topology model employs nodes and links. A node can be a point where two lines intersect, an end point on a line or a given point on a line. For instance, in a road network, the crossing of a township boundary, the end of a cul-de-sac or a tunnel adit may generate a node. A link is a segment of a line between two nodes. Links connect to each other only at nodes.

Theme codes should be taken into consideration when creating nodes to ensure that they are only created between relevant themes, e.g. at the junction between a national highway and a county road and not between roads and property boundaries (unless the road in fact also constitutes a property boundary).

Unique identities are assigned to all links and nodes, and attribute data describing connections are associated with all identities. Topology can therefore be described in three tables:

- The polygon topology table lists the links comprising all polygons, each of which is identified by a number.
- The node topology table lists the links that meet at each node.

- The link topology table lists the nodes on which each link terminates and the polygons on the right and the left of each link, with right and left defined in the direction from a designated start node to a finish node.

A table of point coordinates ties these features to the real world and permits computations of distances, areas, intersections and other numerical parameters. Systems which are based on a topological storage model can be

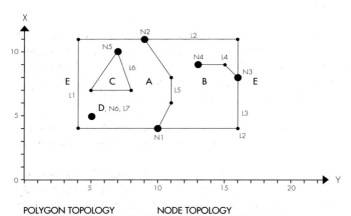

POLYGON TOPOLOGY

Polygon	Links
A	L1, L5
B	L2, L3, L5
C	L6
D	L7
E	L1, L2, L3

NODE TOPOLOGY

Node	Links
N1	L1, L3, L5
N2	L1, L2, L5
N3	L2, L3, L4
N4	L4
N5	L6
N6	L7

LINK TOPOLOGY

Links	Start node	End node	Left polygon	Right polygon
L1	N1	N2	E	A
L2	N2	N3	E	B
L3	N3	N1	E	B
L4	N3	N4	B	B
L5	N2	N1	B	A
L6	N5	N5	A	C
L7	N6	N6	A	A

LINK COORDINATES

Link	Coordinates			
L1	4,10	4,4	11,4	11,9
L2	11,9	11,16	8,16	
L3	8,16	4,16	4,10	
L4	8,16	9,15	9,13	
L5	11,9	8,11	6,11	4,10
L6	10,7	7,8	7,5	10,7
L7	5,5			

4.16 – The idea behind the topology model is that all geometric objects can be represented by nodes and links. The objects' attributes and relationships can be described by storing nodes and links in three tables, – a polygon topology table, a node topology table and a link topology table. An additional table gives the objects' geographical coordinates.

said to be object-based, in the sense that the systems store the geographical objects as identified units. The geometry of the objects is stored in its own subordinate table (see fig. 4.16). Numerous spatial analyses may then be performed, including:

- overlaying
- network analyses
- contiguity analyses
- connectivity analyses

Topological attribute data may be used directly in contiguity analyses and other manipulations with no intervening, time-consuming geometric operations.

Topology requires that all lines should be connected and all polygons closed. Even gaps as small as 0.001 mm may be excessive, so errors should be removed prior to or during the compilation of topological tables.

Topological information permits automatic verification of data consistency in order to detect such errors as the incomplete closing of polygons during the encoding process.

The topological model has a few drawbacks. The computational time required to identify all nodes may be long. Uncertainties and errors may easily arise in connection with the closing of polygons and formation of nodes in complex networks (such as in road interchanges). Operators must solve such problems. When new data are entered and existing data updated, new nodes must be computed and the topology tables brought up to date, which is a major, time-consuming operation.

However, the overall advantages of the topology model over the spaghetti model make it the prime choice in most GIS.

Computer storage techniques and problems are discussed further in Chapter 8. Here, suffice to say that, usually, map data are not stored in a contiguous unit, but rather divided into lesser units that are stored according to a selected structure. This structure may be completely transparent to the user, but its effects, such as rapid screen presentation of a magnified portion of a map, are readily observable.

4.3   RASTER DATA MODELS
4.3.1   GENERAL

Raster data are applied in at least four ways, in:

- models describing the real world
- digital image scans of existing maps
- compiling digital satellite images/data
- automatic drawing driven by raster output units

In the first example, raster data are associated with selected data models of the real world. In the second and third, with compilation methods and in the fourth, with presentation methods. The respective computer manipulations may have much in common; as discussed later, satellite data may be entered in a raster model. For ease of explanation, models based on raster data are discussed below, digital raster maps are treated under "Scanning" in Chapter 6, satellite data are discussed under "Remote Sensing" in Chapter 6, and output units are discussed in Chapter 5.

Model Raster model

The raster model represents reality through selected surfaces arranged in a regular pattern. Reality is thus generalised in terms of uniform, regular cells, which are usually rectangular or square, but may be triangular or hexagonal. Because squares or rectangles are often used and a pictorial view of them resembles a classic grid of squares, it is sometimes called the grid model.

The geometric resolution of the model depends on the size of the cells. Common sizes are 10 X 10 metres, 100 X 100 metres, 1 X 1 kilometre and 10 X 10 kilometres. Many countries have set up national digital elevation models based on 100 X 100 metre cells. Within each cell, the terrain is generalised to be a flat surface of constant elevation.

The rectangular raster cells, usually of a uniform size throughout a model, effect final drawings in two ways. First, lines that are continuous and smooth in a vector model will become jagged, with the jag size corresponding to cell size. Secondly, resolution is constant: regions with few variations are as detailed as those with major variations and vice versa.

The cells of a model are given in a sequence determined by a hierarchy of rows and columns in a matrix, with numbering usually starting from the upper left corner. The geometric location of a cell, and hence of the object it represents, is stated in terms of its row and column numbers. This identification corresponds to the directional coordinates of the vector model. The cells are

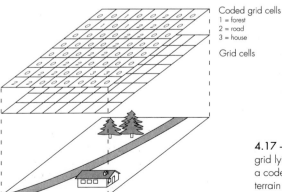

Coded grid cells
1 = forest
2 = road
3 = house

Grid cells

4.17 – Raster data can be visualised as a grid lying over the terrain. Each grid cell has a code stored in the data base describing the terrain within that particular cell.

often called pixels (picture elements), a term borrowed from the video screen technology used in television and computer displays. A pixel is the smallest element of an image that can be processed and displayed individually.

The raster techniques used in GIS are siblings of the rasters long used to facilitate the manipulation and display of information and consequently suit computerised techniques.

Realising the raster model

Raster models are created by assigning real-world values to pixels. The assigned values comprise the attributes of the objects that the cells represent – and because the cells themselves are in a raster, only the assigned values are stored. Values, usually alphanumeric, should be assigned to all the pixels in a raster. Otherwise there is little purpose in drawing empty rows and columns in a raster.

Consider a grid of cells superimposed on the ground or on a map. Assigning the values/codes of the underlying objects/features to the cells creates the model. The approach is comprehensive because everything covered by the raster is included in the model. "Draping" a ground surface in this way regards the ground or map as a plane surface.

Some GIS can manipulate both numerical values and text values (such as types of vegetation). Hence cell values may represent numerous phenomena, including:

- Physical variables such as precipitation and topography respectively, with amounts and elevations assigned to the cells.
- Administrative regions, with codes for urban districts, statistical units, etc.
- Land use, with cell values from a classification system.

Cell table

Cell no.	Cell value	Code list	
00	0	0	un-maped
01	0	1	forest
02	1	2	road
03	0	3	house
04	0		
05	2		
06	0		
.	0		
65	3		
66	3		
67	0		
.	.		
70	2		
71	2		
.	.		

	0	1	2	3	4	5	6	7
0	0	0	1	0	0	2	0	0
1	0	1	0	1	2	2	0	0
2	0	0	0	0	2	2	0	0
3	0	0	0	0	2	0	0	0
4	0	0	0	2	2	0	0	0
5	0	0	2	2	0	0	0	0
6	0	2	2	0	0	3	3	0
7	2	2	0	0	0	0	0	0

4.18 – A line number and column number define the cell's position in the raster data. The data is then stored in a table giving the attribute value of each cell.

- References to tables of information pertaining to the area(s) the cells cover, such as references to attribute tables.
- Distances from a given object.
- Emitted and/or reflected energy as a function of wavelength – satellite data.

A single cell may be assigned only one value, so dissimilar objects and their values must be assigned to different raster layers, each of which deals with one thematic topic. Hence in raster models, as in vector models, there are thematic layers for topography, water supply systems, land use, soil type, etc. However, because of the differences in the way attribute information is manipulated, raster models usually have more layers than vector models. In a vector model, attributes are assigned directly to objects. For instance, a pH value might be assigned directly to the object "lake." In a raster model, the equivalent assignment requires one thematic layer for the "lake", in which cells are assigned to the lake in question, and a second thematic layer for the pixels carrying the pH values. Raster databases, therefore, may contain hundreds of thematic layers.

In practice, a single cell may cover parts of two or more objects or values. Normally, the value assigned is that of the object taking up the greater part of the cell's area, or of the object at the middle of the cell, or that of an average computed for the whole of the pixel.

Cell locations, defined in terms of rows and columns, may be transformed to rectangular ground coordinates, for instance by assigning ground coordinates to the upper left cell of a raster (cell 0,0). If the raster is to be oriented North-South, the columns are aligned along the Northing axis and the rows along the Easting axis. The coordinates of all cell corners and centres can then be computed using the known cell shapes and sizes.

Object relations, which in the vector model are described by topology, are only partly inherent in the raster structure. When the row and column num-

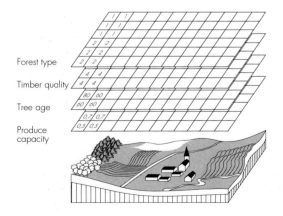

Forest type

Timber quality

Tree age

Produce capacity

Fig. 4.19 – Only one attribute value may be assigned to each cell. Objects which have several attributes are therefore represented with a number of raster layers, one for each attribute.

bers of a cell are known, the locations of neighbouring cells can easily be calculated. In the same way, cells contained in a given polygon may be located simply by searching with a stipulated value. Polygon areas are determined merely by adding up constituent cells. Some operations, though, are more cumbersome. An example of this is the computation of a polygon's perimeter length, which requires a search for, and identification of, all the cells along the polygon's border.

Overviews of phenomena in a given area are easily and rapidly obtained from a raster model by searching all the thematic layers for cells with the same row and column numbers. The relevant overlay analysis is described in Chapter 9.

Coding raster data

Numerical codes and, in some systems, text codes may be assigned to cells. Cell values are entered from word processing files, data bases or other sources in the same sequence as they are registered. See Fig. 4.20.

The way in which the figures are read is dictated by format. For instance, it is essential to know the number of columns per row.

Raster data may be available from a variety of dissimilar sources, ranging from satellite data and data entered manually to digital elevation data. Their collocation requires that cells from differing sources and thematic layers correspond with each other. In other words, cells having the same row and column numbers must refer to the same ground area. Various computations may be necessary in order to accommodate any differences in cell shape and size.

Cells may contain values referenced to attribute tables. The cells of a thematic layer may be coded so that their values correspond to identities in a given attribute table.

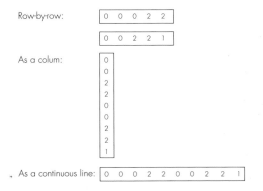

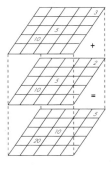

Fig. 4.20 – Typical cell input. Raster data may be enterred in the data base as a series of rows, a series of columns or as a continuous line.

4.21 – It is possible to make direct calculations on the raster layers.

Attribute data or tabular data may be coded independently, irrespective of whether the geometry is represented using vector data or raster data.

4.3.2   STORING RASTER DATA

If the cell values of a raster model are entered in fixed matrices with rows and columns identical to those of the registered data, then only the cell values need to be stored; row and column numbers need not.

Even when only the cell values are stored, the volumes of data can easily become unwieldy. Typical operations may involve 200 thematic layers, each containing 5,000 cells. The total number of cell values stored is thus 200 X 5,000 = 1 million. A Landsat satellite raster image contains about seven million pixels, a Landsat TM image about 35 million pixels.

Various devices may be employed to reduce data volume and, as a consequence, storage memory requirements. Cells of the same value are often neighbours because they pertain to the same soil type, the same population density of an area, or to other similar parameters. Thus, cells of the same value in a row may be compacted by stating the value and their total. This type of compacting is called run-length encoding and is illustrated in Fig. 4.22. Further compacting may be achieved by recursively applying the same process to subsequent lines.

The volume of the data also depends on the size of the cells. If, for instance, the cell size is reduced from 100 m X 100 m to 50 m X 50 m, the total number of cells quadruples. Conversely, increasing the cell size to 200 m X 200 m would quarter the total number of cells. In other words, in representing a given area, the aggregate amount of data involved is proportional to the square of the resolution (into cells).

This relationship may be employed to save storage space. Larger cells (lesser resolution) may be used to represent larger homogeneous areas, whilst

4.22 – Run-length encoding is the method used to save data storage space by reducing a row of cells with the same value to a single unit having a specific value and quantity.

4.23 – Computer memory capacity requirements can be reduced by storing homogeneous and uniform (quadratic) areas, consisting of many basis cells, as one single unit.

smaller cells (greater resolution) may be used for more finely detailed areas. The approach is known as the quadtree representation and is illustrated in Fig. 4.23.

Computer storage techniques are discussed further in Chapter 8.

4.4 AUTOMATIC CONVERSION BETWEEN VECTOR AND RASTER MODELS

GIS applications sometimes require data in a form differing from that which is available. As a result, many GIS now have facilities for automatic conversion between vector and raster models. Raster data are converted to vector data through vectorisation. The reverse process, which is just as common, is rasterisation.

In vectorisation, areas containing the same cell values are converted to polygons with attribute values equivalent to the pre-conversion cell values. In the reverse process of converting polygons to cells, each cell falling within a polygon is assigned a value equal to the polygon attribute value.

Various routines are available for converting raster data to vector data and vice versa. Converting raster data to vector data is the more complex and time consuming of the two processes, and different conversion programs can yield

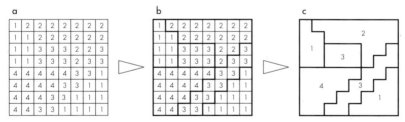

Fig. 4.24 – Conversion of raster data to vector data. a) Each raster cell is assigned an attribute value. b) Boundaries are set up between different attribute classes. c) A polygon is created by storing x- and y-coordinates for the points adjacent to the boundaries.

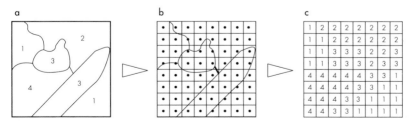

4.25 – Conversion of vector data to raster data. a) Coded polygons: b) A grid with the right cell size overlays the polygons. The polygons which contain the centre of the individual cells are identified, c) Each cell is assigned the attribute code of the polygon to which it belongs.

differing results from the same set of raster data. In all cases, some information/data are lost in all conversions. Consequently, converted data is always less accurate than original data.

These conversion processes apply specifically to data, not to the conversion of raster data from scanned maps into vector form, which is discussed further in Chapter 6.

4.5   VECTOR VS. RASTER MODELS

One of the basic decisions in GIS design involves the choice between vector and raster models, each of which has advantages and disadvantages. In the vector model, the observation units are end points and/or variable line or polygon magnitudes, whereas the raster model presupposes fixed observation areas in a grid. Otherwise the models are identical.

Many real phenomena are related to locations. Measurements are often made at points, infrastructures are often related to lines, and administrative units are frequently described in terms of defined areas of various shapes and sizes.

Raster GIS emphasises properties: here, the basic units of observation are regular cells in a raster. Not all phenomena are related directly to such grid patterns. At the time of writing (late 1991), satellite data and digital elevation data account for the bulk of data available in raster form. Other types of data usually must, to a greater or lesser degree, be reworked to suit rasters. Raster data accessibility may thus be a major problem and perhaps the greatest drawback of a raster GIS in comparison with a vector GIS.

On the other hand, a vector model often requires the time consuming and costly compilation of digital map data, whilst maps are integral parts of the data compiled for a raster model. Maps may be drawn for all cells as soon as they are assigned values.

Despite over-simplification from a functional viewpoint, vector data may be considered best suited for documentation whilst raster data are more adept at showing the geographic variation of phenomena. Another simplification might be that vector data are preferable for line presentations while raster data are superior for area presentations. Selected characteristics of the vector and raster models are compared in Fig. 4.26.

To date (late 1991), the vector model has been dominant in commercial GIS

	Raster	Vector
Data collection	Rapid	Slow
Data volume	Large	Small
Graphic treatment	Average	Good
Data structure	Simple	Complex
Geometrical accuracy	Low	High
Analysis in network	Poor	Good
Area analysis	Good	Average
Generalization	Simple	Complex

Fig. 4.26 – This table illustrates some of the characteristic differences between raster and vector data.

implementations. The raster model, on the other hand, has been used more frequently in natural resource planning and management and in teaching because it is more easily explained and used.

Many newer GIS can manipulate both vector models and raster models. With dual capability, a GIS can exploit the respective advantages of both models: vector data might be converted to raster data to perform overlaying or other operations more easily performed using rasters, and then converted back to vector data.

4.6 TABULAR DATA AND COMPUTER REGISTERS

With the advantages of easy updating, rapid search and the flexible superimposition of data, the computerised filing of information has become commonplace in administrative work. Frequently inaccessible, massive quantities of traditional records and files have been replaced by work stations from which enormous amounts of information are rapidly accessible. Physical separation by rooms, buildings, national borders or intervening distances is no longer a barrier to ready availability of information.

In the days when all records were kept on paper, each agency, organisation or user tended to structure their own manual files. The result was a proliferation of parallel files, often containing nearly identical material. Computerisation permits a simplification and coordination of registration efforts and can eliminate duplication and rationalise the overall filing process. In the public sector, central registers have been established as a common resource for numerous users.

Most countries have various computerised registers, including registers of:

Fig. 4.27 – Manual and electronic files. Manual files require considerable storage space and are difficult to access. Electronic files, on the other hand, are compact and easily accessible, although relatively expensive.

- customers/clients
- motor vehicles
- properties
- literature
- moveables
- buildings
- addresses

- citizens
- pleasure boats
- deeds
- environment
- companies
- roads and signs
- personal data

Some of these registers are important in GIS applications. Others are of lesser interest. In many countries, though, work is under way to make public registers available to GIS users.

Upon entry, register data are selected (structured) so that registers contain uniform and limited data. As for digital map data, register data are stored using formats.

There is no one general pattern for register content, but usually the items registered will have identities, locations (addresses, etc.), descriptive details, time and date notations, and sometimes references to other registers.

For example, typical entries in a building register may contain details listed in Fig. 4.29.

GIS users may be obliged to compile registers when they are unavailable in computerised form, either by using GIS techniques directly or by using other systems and subsequently exporting data to GIS. In all cases, data models should be specified.

Georeferencing
ID
Geometry/ Coordinates
Topology

Attributes
ID
Location
Description
Dating

Fig. 4.28 – Geometric content is often limited to I.D., geometry/coordinates and topology; whereas attribute content often comprises I.D., location e.g. address, various descriptions of the object and dating (age etc.).

BUILDING REGISTER	
Identity:	• Building number
Location:	• Address • Representative coordinates
Description:	• Builder/owner • Status (under construction, in use, demolished) • Type (detached house, semi-detached house, farmhouse, garage) • Function (residence, farm, shop) • Structure (concrete, frame, brick) • Water supply (public, cistern, stream) • Sewage (public, subsoil disposal, closed system) • Heating (electric, oil, wood) • Available area
Dating:	• Year built
References:	• Reference to property register • Reference to address register

Fig. 4.29 – Typical contents of a Building Register.

Coding and entering attribute data

Attribute data may be coded for several reasons in order to:

1. Establish an i.d. code between geometry and attributes.
2. Conserve computer memory.
3. Ease input work.
4. Ease verification of data entered.
5. Simplify subsequent searches for data in databases.

The coding of geographic data is not new. Systems have been established in many fields for coding pipes, manholes, streets, properties, buildings, the names of towns and so on. Indeed, codes have been used for many reasons, not least as file access keys or to conserve the used on file cards.

Codes are often assigned according to a hierarchical classification system devised to ease such data operations as searching and sorting. Examples include the official codes widely used for addresses, names of towns, highways, etc.

For example, a land use application might involve a hierarchy such as that illustrated in Fig. 4.30.

Attribute data may also be tabulated, as illustrated in Fig. 4.31.

Tabulation eases sorting, such as finding all wooded objects or all transportation objects by using their respective land use category codes.

The column headings in the table are the names of attribute types. The type of data may be specified for each field, such as integer (land use code), decimal (area), text (name), etc. Code tables may be compiled and used with the main table to produce more meaningful printouts from the system.

Attribute data may be entered relatively easily in most GIS, either manually via a keyboard or by importing data from an existing register; see

Level 1 Code	Attribute	Level 2 Code	Attribute	Level 3 Code	Attribute
100	Built-up	110	Industry	111	Light
				112	Heavy
				113	Other
		120	Transportation	121	Roads
				122	Railways
				123	Airports
				124	Parking
				125	Terminal
				126	Other
		130	....	131	....
120	Wooded	210	Coniferous	211	Fir
				212	Pine
				213	Ohter
		220	Deciduous	221	Oak
				222	Beech
				223	Other
		230	....	231	....

Fig. 4.30 – Land use code hierarchy. Attributes are often coded in accordance with an hierarchical classification system.

ID code	Land use code	Area ha	Township code
1	222	22.67	0914
2	211	1.45	0916
3	121	46.80	0923

Fig. 4.31 – Tabulated attribute data. Attribute data can be sorted and arranged in different tables.

Chapter 6 for further details. I.d. codes are usually entered together with the attribute data. They may also be registered or edited into compilations of attribute data which initially have no codes.

Storing attribute data

Attribute data are usually most easily and expediently stored in tabular form. Each line in a table represents an object, each column an attribute. Attribute data are therefore often called tabular data and are normally stored in a relational data base; see Chapter 8.

Data on different types of objects are usually stored in separate tables, each dedicated to a single object type. In each table, line formats and lengths are identical throughout. The number of columns may be extended by combining several tables, either by using a common access key or by manually entering new attributes.

In principle, table design is independent of whether the geometrical data to which attributes refer is in the form of vector data or raster data. However, table content must be relevant to the objects, so each object/line must have a stable identity or access key.

Computerised registers are not always compiled with equal ease. Consider the common example of road data which are often referenced to distances from intersections or roadside markers. Whenever a road is rerouted, distan-

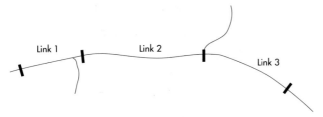

ATTRIBUTE TABLE

Link ID.		Distance	Road classification	Cover	Speed limit
1	Hp 1	0 – 800	Rv. 12	Asphalt	80
2	Hp 1	800 – 1200	Rv. 12	Asphalt	60
3	Hp 2	0 – 900	Rv. 12	Gravel	60

Fig. 4.32 – Road links as carriers (bearers) of information.
The reference system often used for defining attribute data linked to roads is based on the measurement of distances from a zero-point.

ces change. This, of course, changes the reference system. Furthermore, links are customarily selected to carry information, which may compound the task because data volume and complexity increase in proportion to the number of links. The first step, how a road is divided into links, determines the nature of the nodes. If all intersections, events and features along the road result in nodes, the number of links may be enormous

Data available in existing computerised registers is not always in convenient tabular form. As a result, conversions and roundabout methods must often be used to access such data for GIS uses; see Chapter 6.

4.7 LINKING DIGITAL MAP AND REGISTER INFORMATION

Common identifiers in map data and attribute data permit moving from map data to attribute data and vice versa. Attribute data which lacks georeferencing may be linked to geography. As illustrated in Fig. 4.33, this is possible if the attribute data which lacks georeferencing have access keys in common with attribute data that have other access keys in common with map data. The connection is then from attribute data to other attribute data to map data.

This illustrates one of the distinctive capabilities of GIS. Data which initially contain no geographic information or referencing may be given geographic dimensions and may therefore be used to enhance and present data in new ways, in maps or on screen.

DIGITAL MAP DATABASE

Point no.	Theme code	Building	X-Coordinate	Y-Coordinate
1567	10	589	10,112,00	6,771,00

BUILDING REGISTER

Building no.	Lot. no.	Owner	Year built	Type of building
589	44/113	Peder Ås	1952	House

LAND REGISTER

Lot. no.	Landowner	Date of birth	Area m²	Property address
44/113	Nils Nilsen	200519 291 89	800	Toppen 1, 1099 Fjel

PERSONAL REGISTER

Date of birth	Name	Sex	Address
200519 291 89	Nils Nilsen	male	Svingen 3, 4999 Vika

Fig. 4.33 – The figure above illustrates how elements of data in one register can be used as an access key to another register - thereby acting as a link between dissimilar data. A pre-condition is that identical information must be available in a minimum of two registers.

Least common map unit

Map data may be used not just to link maps and attributes, but also geographically to link dissimilar attributes.

Superimposing dissimilar data, such as geological data and vegetation data, is often hampered by a lack of commonality between the observations made in the field. That is, the observation areas listed in the respective attribute tables cannot be listed together because they refer to different sets of locations. In GIS this problem can be solved by using cartographic integration, in which overlay techniques (see Chapter 9) are used to combine geometries from two dissimilar thematic maps into a single synthesised map. The synthesised map contains numerous new objects and areas, all of which are related to the two original thematic maps. Hence the objects in the synthesised map comprise the least common units between the original maps, and are therefore called Minimal Mapping Units (MMU).

An attribute table is associated with the MMUs. In it, the MMUs are listed on the lines, and the elements of the original thematic maps are in the columns. This table contains all the relevant attributes and therefore may be used in further analyses of the data.

Cartographic integration is straightforward when the areas of the original map data contain more or less homogeneous data, such as property data, land use, vegetation, geology, etc. Complexities arise when properties are not evenly distributed over area. Consider, for instance, a typical township with an unevenly distributed population that averages 100 persons/km^2. An MMU might locate in an uninhabited area of the township and hence misrepresent

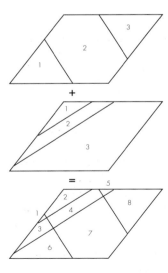

ID.	Attribute
1	Cultivated
2	Grazing
3	Forest

+

ID.	Attribute
1	Development
2	Recreation
3	Preservation

=

ID.		Attribute		Attribute
1	1	Cultivated	1	Development
2	2	Grazing	1	Development
3	1	Cultivated	2	Recreation
4	2	Grazing	2	Recreation
5	3	Forest	2	Recreation
6	1	Cultivated	3	Preservation
7	2	Grazing	3	Preservation
8	3	Forest	3	Preservation

Fig. 4.34 – Attribute data can be rendered comparable by superimposing geometries from dissimilar geographic units over each other to find "Minimal Mapping Units".

the facts of its population. In all such cases, rules must be contrived to designate how attributes shall be divided among MMUs.

4.8 GEOMETRIC DESCRIPTION OF CONTINUOUS TERRAIN SURFACES – DIGITAL TERRAIN MODELS

The digital representation of a terrain surface is called a Digital Terrain Model (DTM) or a Digital Elevation Model (DEM).

In GIS disciplines, the term Digital Terrain Model is often used not just for the model itself, but also for the software used to manipulate the relevant data. Only the model is discussed in this chapter; data manipulation is primarily discussed in Chapter 9.

Terrain surfaces are continuous surfaces with continuously varying relief. Prominent terrain features are described verbally in many terms, including smooth slope, cliff, spot depression, spot height, characteristic line (such as the top or bottom of a road cut), saddle and so on. Geometry, however, has only three terms: point, line and area. In principle, one cannot describe continuously varying terrain using only three discrete variables, so all descriptions are necessarily approximations to reality.

Essentially, digital terrain models comprise various arrangements of individual points in x-y-z coordinates. Often their purpose is to compute new spot heights from the originals.

Point model

A systematic grid, or raster, of spot heights at fixed mutual spacings is often used to describe terrain. Elevation is assumed constant within each cell of the grid, so small cells detail terrain more accurately than large cells. The size of cells is usually constant in a model, so areas with a greater variation of terrain may be described less accurately than those with less variation.

Data on individual points or lines may be used to form the grid. When the data points are dispersed, the averages of the elevations of those closest to grid points, within a given circle or square, are assigned to the grid points with inverse weighting in proportion to the intervening distances involved. When the data relates to profiles or contours, grid point elevations are interpolated from the elevations at the intersections of the original data lines, and the lines of the grid.

Terrain may also be described in terms of chosen or arbitrarily selected individual points (point cloud). In principle, the characteristics of the terrain between points are unknown, so it follows that point densities should be greatest in areas where terrain features vary the most.

Fig. 4.35 – Digital terrain models describe the terrain numerically in the form of x-, y- and z- coordinates. Graphic presentation can be either in the form of a grid (as in the figure) or as profiles.

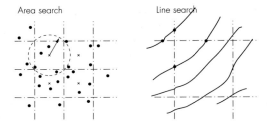

Fig. 4.36 – The DTM-based grid model is created either from an area search in a point cloud, or from intersections with contour lines.

Only elevations are stored for the points of a regular grid, but both point coordinates and elevations must be stored for point clouds. So, for given terrain coverage, the amount of memory storage required for the two point arrangements differs. In describing abrupt terrain variations, such as the top and bottom of a cut, point models are inferior.

Line model

Isolines – continuous lines connecting points of the same elevation – may represent terrain in much the same way as contour lines depict terrain on conventional maps. The point densities should be greatest in those areas in which the terrain varies the most. As the intervening terrain between successive isolines is unknown, smaller elevation increments between isolines result in greater accuracy of description.

Although an isoline model may be compiled readily, amending its data is involved. In practice, the methods used are determined by the data compilations.

Parallel profile lines, connecting points of varying elevation, may be used to describe terrain. The density of points along profile lines should be increased in areas where there are major variations in the terrain. In principle, the terrain between successive profile lines is unknown so the closer the lines, the greater the accuracy of description.

A combination of isolines and individual points may also be used to describe terrain, especially when specifying such point features as peaks and valley floors or vital terrain lines, such as the top and bottom of a fill.

Area model (TIN)

An area model is an array of triangular areas with their corners stationed at selected points of great importance, for which the elevations are known. The inclination of the terrain is assumed to be constant within each triangle. The area of the triangles may vary, with the smallest representing those areas in which the terrain varies the most. The resultant model is called the Triangulated Irregular Network (TIN).

Fig. 4.37 – The surface of the terrain can be described as inclining triangular areas, starting from selected points of importance. This method of describing terrain is called the "Triangular Irregular Network" (TIN).

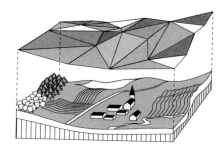

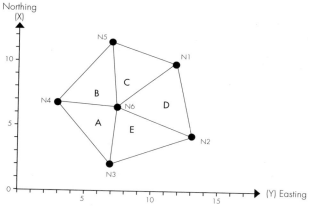

Triangle table

Triangle	Adjacent
A	B, E
B	A, C
C	B, D
D	C, E
E	D, A

Trinangle/Node Table

Triangle	Nodes
A	N3, N4, N6
B	N4, N5, N6
C	N1, N5, N6
D	N1, N2, N6
E	N2, N3, N6

Coordinate tables

Node	Coordinates
N1	X_1, Y_1, Z_1
N2	X_2, Y_2, Z_2
N3	X_3, Y_3, Z_3
N4	X_4, Y_4, Z_4
N5	X_5, Y_5, Z_5
N6	X_6, Y_6, Z_6

Fig. 4.38 – In the area model (TIN), the triangles are stored in a topological structure.

Insofar as possible, small equilateral triangles are preferable. In the TIN model, the x-y-z coordinates of all points, as well as the triangle attributes of inclination and direction, are stored. The topological data storage structure comprises polygons and nodes.

Compared to the grid model, the TIN model is relatively cumbersome to establish but more efficient to store because areas of terrain with little detail are described with less data than similar areas with greater variation.

Practical observations

Grid models and area models are always less accurate than the original data from which they are derived. Therefore, in some GIS/DTM, original data are stored in point clouds. Models with accuracies suiting specific tasks are compiled from these as required. For instance, a grid model for estimating road construction excavations might be more accurate than one intended to detail the general vegetation of the region.

GIS based on vector models can easily manipulate elevations stored as spaghetti data, but can less easily handle elevation grid data. Only a GIS based on a topological model can manipulate TIN data.

Terrain data are usually compiled from survey point elevations, from isolines digitised from existing maps, or from photogrammetric point and/or line (contour or profile) registration. Normally photogrammetric point registration is 30% more accurate than line registration.

Various interpolation programs compute new z coordinates for new x-y coordinates, thus facilitating specific computations such as estimating cut and

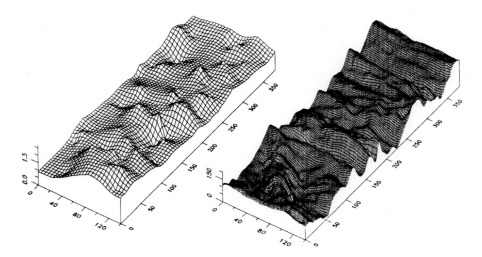

Fig. 4.39 – By placing different themes on the z-axis, the distribution of varying amounts over the terrain can be illustrated. This figure shows: a) Differences in elevation in an area, b) biomass production in the same area. (Smith, Sheider, Wiart, 1987)

fill volumes in road planning, or assessing reservoir volumes for hydroelectric plants. Various GIS may also be implemented with functions for calculating slopes, drawing in perspective in order to visualise the impact of works, computing the runoff or, perhaps, draping in colours to enhance visualisation. See Chapter 9.

The ways in which data are represented and stored are decisive in determining the type and efficiency of the computations. For instance, digital isolines are ill-suited to calculating slopes and relief shadowing. Draping and runoff calculations are most expediently performed using a TIN model. The TIN model is ill-suited to visualisation without draping, and so on.

The methods used to describe terrain surfaces may also be used to describe other continuously varying phenomena. Thus, population density, prevailing temperatures or biomass production can be described simply by assigning the parameter involved to the z axis of all the observation points located in x-y coordinates.

4.9   GEOREFERENCES AND COORDINATE SYSTEMS

Before dissimilar geometrical data may be used in GIS, they must be referenced in a common system. The problems associated with dissimilar geometrical data are hardly new: there are numerous georeference systems which describe the real world in different ways, and with varying precision.

4.9.1   CONTINUOUS GEOREFERENCE SYSTEMS

Continuous georeferencing implies continuous measurement of the position of phenomena, with no abrupt changes or breaks. It involves resolution and precision. Resolution refers to the smallest increment that a digitiser can detect. It is similar to precision, which is a measure of the dispersion (in statistical terms, the standard deviation). Accuracy, however, is the extent to

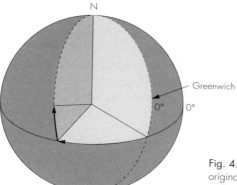

Fig. 4.40 – The geographical coordinate system originates from the earth's centre and defines a location in degrees in relation to the Equator and the zero meridian at Greenwich.

which an estimated value approaches a true value, i.e., the degree to which it is free from bias.

Theoretically, there is no limit to how precise continuous measurements may be. Distances may, for instance, be measured with a precision of 1 kilometre, 1 millimetre, or 1 nanometre. Many geographical phenomena are measured on a continuous basis, including property boundaries, manhole locations, building details and, of course, many map details.

Continuous systems involve:
- coordinates on the curved surface of the Earth
- rectangular coordinates
- geocentric coordinates

The geographical coordinates on the surface of the Earth are the familiar latitude, measured in degrees north or south of the equator, and longitude, measured in degrees east or west of Greenwich in England. Positions in latitude and longitude are relative only. Distances and areas must be calculated using spherical geometry and the Earth's radii to the points in question.

In applications, latitude and longitude are usually used in describing major land areas.

Georeferenced data may be drawn on maps only when referenced to a plane surface, not to the curved surface of the Earth. Therefore, many countries have national (and local) georeference systems of rectangular coordinates which permit locations to be given in units of length relative to a selected origin. Most systems comprise x and y axes and coordinates. The coordinate system orientation may differ, so coordinates should always be identified unambiguously, for example, in terms of compass directions (northing, easting) from the origin. Furthermore, coordinate systems usually include elevations relative to a datum, usually mean sea level.

Various projections are used to represent the curved surface of the earth on the plane surface of a map. They classify in three groups according to the underlying geometrical transformations involved, as cylindrical (Mercator and UTM Grid – Universal Transverse Mercator Grid, etc.), conical and azimuthal.

Ordinary geometrical computations may be made in rectangular coordinates but it should be remembered that all projections incur errors in distance, area, direction or shape, and that these errors rise with the increasing size of the area represented. This difficulty is most often overcome by constraining axis systems to relatively small areas. Larger areas are then represented by several axis systems, displaced with respect to each other. In areas where neighbouring axis systems overlap, one must often work in two systems.

Many GIS have facilities for transforming data from one coordinate system to another.

In some continuous georeference systems, objects are located in direction and distance from specified objects in terrain. For instance, the locations of buried pipes are often given relative to terrain features such as road edges, buildings, poles etc. Such measurements can seldom be translated directly to existing rectangular coordinate systems.

Geocentric coordinates are based on a rectangular coordinate system with an origin at the centre of the Earth. The z axis is coaxial with the axis of the Earth's rotation and is positive in the direction of the North Pole. The x axis goes through the zero meridian in Greenwich; the y axis is orthogonal to the left of the positive x axis, as shown in Fig. 4.43.

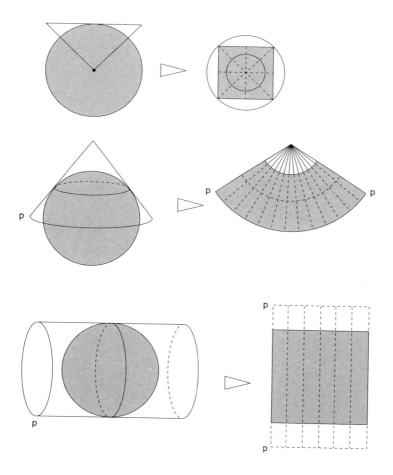

Fig. 4.41 – Map projections. a) azimuthal projection, b) conical projection, c) cylindrical projection.

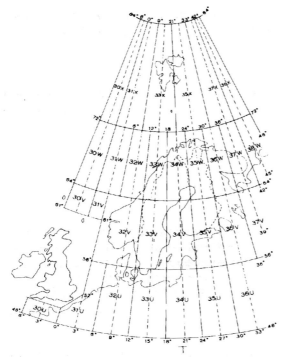

Fig. 4.42 – UTM (Universal Transverse Mercator).
To minimise geometrical errors in map projections, cylindrical
projections are placed closed together so that only relatively
small areas overlap each tangent axis.

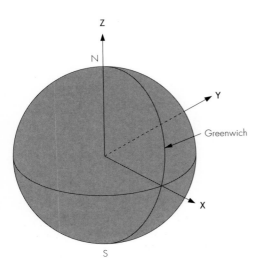

Fig. 4.43 – The geocentric coordinate
system is rectangular, with an origin at the
centre of the earth and axes as shown.

The advantage of the geocentric coordinate system is that it covers the entire Earth, which is why it is used for GPS georeferences, see Chapter 6.

4.9.2   DISCRETE GEOREFERENCE SYSTEMS

In discrete georeference systems the positions of phenomena are measured relative to fixed and limited units of the surface of the Earth. Typical reference units include:

- address and street codes
- postal codes
- statistical units and other administrative zones
- grids

The unit size determines the accuracy of registration: the smaller the units, the greater the accuracy. Discrete systems are often based on code indices with no inherent usefulness in map representations. So, save for grids, they are ill suited for GIS applications. Nevertheless, the data registered may be linked to rectangular coordinates through reference transformations or through cartographic fixing of reference units in a rectangular coordinate system.

Many geographic phenomena are registered in discrete systems. Address and street codes, for example, may be used to indicate locations along roads. Roads and streets are then divided up into selected registration units, such as city blocks, or fixed or variable distances. Postal codes, like statistical and administrative units, locate phenomena within geographic districts of variable shape and size. The relevant geographic districts are often arranged in a hierarchy.

Fig. 4.44 – Discrete location registration may be based on a reference system of administrative or statistical geographical units.

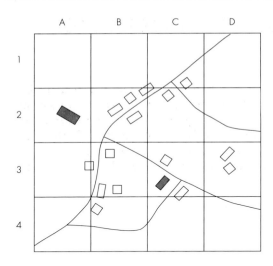

4.45 – A typical example of a discrete reference system is the index system common on city maps.

Index

Town hall	2A
Sports center	3C

Grids give positions in the same way as grids used in GIS. Typical examples include the indices common on city maps or on national sheet map survey series.

Discrete reference systems are often easy to use and are therefore expedient when accuracy is not a major concern. For instance, finding the street code of a new house is undoubtedly easier than locating the exact coordinates of its four corners. However, problems often arise whenever boundaries are moved, changing unit shapes and sizes.

5. HARDWARE AND SOFTWARE FOR GIS

GIS tasks are many and varied, so no one system assembled from available computers, peripheral equipment and software can be expected to meet all current and future needs. Moreover, the variety of computer hardware, software, quality and prices on the market is broad and expanding. So matching a system to defined needs remains crucial in implementing GIS facilities.

As described below, the physical parts of a GIS facility collect, store, manipulate and present data.

5.1   HARDWARE
5.1.1   COMPUTERS

The heart of a modern computer consists of one or more integrated-circuit microprocessors, or Central Processing Units (CPU), which contain the circuitry necessary to interpret and execute program instructions for processing data and controlling peripheral equipment.

All data are stored in a working memory together with the operating system and the application program, which may be entered from secondary storage as needed. The working memory is completely solid-state electronic and therefore access to, and retrieval from it, are rapid. The total storage capacity is usually 1 to 10 Megabytes. A byte comprises eight adjacent bits, processed as a unit and capable of holding one character or symbol. Mega is a prefix normally indicating a magnitude of one million in the decimal system. However, as shown in Fig. 5.2, in the binary system used in computers it indicates a magnitude of two to the twentieth power.

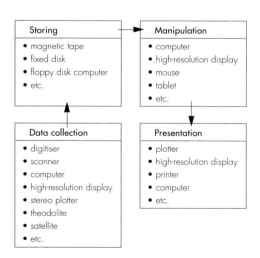

Fig. 5.1 – A geographic information system can comprise equipment such as that shown in the illustration. This equipment is intended to meet requirements for data collection, storage, manipulation and presentation.

Prefix	Meaning in decimal system Example: frequency in Hertz	Meaning in binary system Example: capacity in bytes
kilo	$10^3 = 1,000$ Hertz kilohertz	$2^{10} = 1,024$ bytes (kHz) kilobyte (kbyte)
mega	$10^6 = 1,000,000$ Hertz megahertz (MHz)	$2^{20} = 1,048,576$ bytes megabyte (Mbyte)
giga	$10^9 = 1,000,000,000$ Hertz gigahertz (GHz)	$2^{30} = 1,073,741,824$ bytes gigabyte (Gbyte)

Fig. 5.2 – Common prefixes in decimal and binary systems.

The secondary storage is used to store permanently large quantities of data, such as GIS data that is not actively being processed. Secondary storage may be:

- fixed and removable disks with capacities from 20 megabytes (in PCs) up to several hundred gigabytes
- floppy disks with a capacity of 360 kbytes to 1.44 megabytes
- magnetic tapes of 10 megabytes or more
- optical disks of 250 megabyte or more

A secondary storage data search is mechanical and therefore slower than a working memory search. Magnetic tape searchs are particularly slow, especially if the data sought are stored towards the end of the tape. The current trend is towards faster, larger and cheaper storage devices.

The operating system is a basic program which administers the internal data processing in a computer. Different operating systems (MS-DOS, OS/2, UNIX, VMS, etc.) are used in different types and makes of computer. So, as application programs cannot be moved from one operating system to another without extensive modification, the choice of application program is often determined by the choice of computer.

However, current trends are broadening the alternatives and many suppliers of GIS programs now offer versions matched to different operating systems. Computer manufacturers are increasingly offering wider ranges of equipment for the operating systems, such as UNIX in work stations and MS-DOS in PCs.

Three types of computer may be used in GIS:

- the microcomputer and two types of smaller computer, the personal computer (PC) and the work station. Originally, the personal computer was designed for individual use, whilst the work station was intended to provide users with access to a larger computer. However, developments in the late 1980s and early 1990s diminished the differences between the two types because PCs are now used in networks and work stations can be single units for individual use. The only remaining difference is that PCs use operating systems developed for PC use, whilst work stations use operating systems originally developed for larger computers.

- minicomputer: a smaller mainframe computer.
- mainframe: a large computer, usually the hub of a system serving several users.

The types and divisions between the two are continually changing.

PCs and work stations are now the types used most frequently for GIS applications. As shown in Fig. 5.4, PCs are becoming dominant. This trend may be ascribed to the explosive developments in both PC technology and markets during the late 1980s and early 1990s. Most importantly, in terms of processing speed and storage capacity (the two main measures of computer capability), the PCs of the early 1990s outperform the minicomputers of the late 1980s and the mainframes of the late 1970s, at a fraction of the cost. Secondly, the enormous popularity of PCs has made hardware and software products for them readily available at highly competitive prices. Finally, many of the features formerly afforded by minicomputers and mainframes are now available in PC systems. Among the most widespread PC systems are various Local Area Network (LAN) configurations, which facilitate multi-user access without restricting local computing ability. Some of the "traditional" limitations of PCs, such as the 640 kbyte working memory restriction imposed by the ubiquitous MS-DOS system, have been overcome by a variety of software artifices. Finally, computer interconnections are now common, so users need no longer be constrained by the choice of computers within one of the four major categories. Systems now available and operating may include several PCs and workstations or even a mainframe computer connected to a Local Area Network with one or more servers.

Servers are computers dedicated to serving the network and managing an expensive shared resource. There are many types; the more common include a communications server, or gateway, that interfaces to other networks including public telephone networks; a printer server that permits a single high-quality printer to serve all terminals connected to the network; and a file

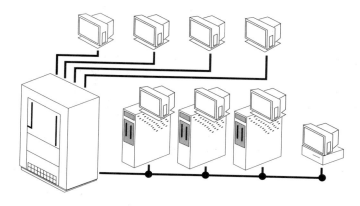

Fig. 5.3 – Typical Local Area Network (LAN).

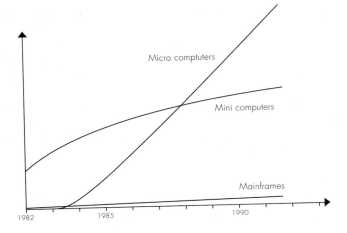

Fig. 5.4 – Trends in computer types used for GIS applications.

server that provides storage capacity beyond that available in the individual terminals.

5.1.2 DISPLAY

The display, or monitor of the computer is the user's prime visual communication medium in all computer work. A display consists of a screen and its associated electronic circuitry. The screen itself is either the face of a cathode ray tube (CRT), similar to those long used in television sets and radars, or one of the newer, flat arrays of semiconductor elements, such as the liquid crystal displays (LCDs) used in "laptop" and "notebook" PCs. At the time of writing (late 1991), LCD technology is advancing, but the CRT remains the mainstay for almost all GIS applications because it produces more light, has higher resolution and a greater range of possible display colours than flat screens.

All CRTs are vacuum tubes/valves. They operate on the common principle that a beam (ray) of electrons emitted by a cathode is focused and directed at an electroluminescent screen. Apart from the physical characteristics of size and colour capability of electroluminescent screen phosphor (monochrome or colour), CRTs divide into two major groups according to their persistence, or how long the displayed information remains on the screen surface. On regeneration screens, information displayed is periodically refreshed. On storage screens, information displayed is retained until changed.

Storage screens are seldom used, if at all, in new equipment. The storage screen technology may be viewed as an interim analogue solution to a problem that is now handled digitally. They were originally devised for oscilloscopes, to retain displayed waveshapes long enough to permit viewing or photographing. The retention function is now performed by storage in a digital memory, so short-duration waveshapes may be displayed on regeneration screens.

Regeneration screen images are refreshed up to 60 times a second and displays are based either on rasters, as with TV images, or on vector tracings of individual lines. Usually, GIS vector map data are converted to raster data for screen presentation.

Colour screens employ the three basic colours – red, green and blue (RGB) – in varying intensities which reproduce a wide range of hues. A high-resolution 19-inch (diagonal) screen designed for computer graphics may resolve 1280 points horizontally and 1024 points vertically.

Screens are now available with processing functions which include vector-raster conversion, storage of vector data and colour Look-Up Tables (LUT) for rapid alteration of colour patterns. High-definition colour screens of the regeneration type are used most frequently in GIS applications.

The on-screen position of the cursor, the movable, sometimes blinking, symbol that indicates the position of the next operator action, is controlled in many ways. On keyboards, this is usually by the four arrow keys and various function keys. Provided the computer involved is suitably equipped, the cursor position may also be controlled by a joystick or a mouse, a palm-sized, button-operated device which, when moved across a flat surface, produces a corresponding movement of the cursor on the screen.

Tablets are also used to control cursors and communicate with application software. A tablet is a small digitiser arranged to accept overlays. By moving a penlike device or a mouse, the cursor on the screen is moved and software instructions are initiated by pointing at fields in the menu.

Some computers support the use of a light pen, which, when its tip is pointed at a feature in the on-screen display, registers its coordinates and permits selective functions to be performed, such as the retrieval of information about the objects pointed out.

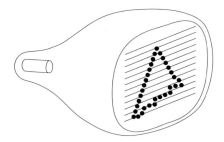

Fig. 5.5 – Fig. a) shows a raster regeneration screen.

Fig. 5.6 – The mouse is used to communicate with application software. By moving the mouse, the "cursor" on the screen is moved. Program commands are activated by pointing and clicking the mouse buttons.

5.1.3 QUANTIZERS

Quantizers are devices that convert analogue data, like data on ordinary maps, into digital data, as used in GIS. The quantizers used most frequently include the:

- Digitiser: a combination of manual positioning and electromagnetic sensing on a plane surface; historically the first analogue-digital converter for GIS applications.
- Scanner: an automatic electromechanical system that converts a picture into a raster of points (pixels).
- Video: the electronic equivalent of motion-picture photography, often followed by rasterisation similar to that employed in a scanner.
- Automatic line tracer: an automated version of the digitiser in which an "intelligent" sensor automatically follows lines on maps and on screens.

Digitisers

A digitiser consists of a flat, non-conducting surface in which is embedded a grid of wires. Electric currents passing through the wires give rise to electromagnetic fields which together form a pattern on the surface. A small magnifying glass fitted with cross hairs is used to locate objects on a map placed on the table. The glass is also fitted with a sensor coil, which detects the position of the cross hair centre with respect to the underlying wire grid. Thus, positions are entered continuously or at the command of a keystroke. A keypad connected to a microprocessor allows up to 25 codes (identifiers, thematic codes etc.) for objects that have been located to be entered, and also facilitates interaction with the program involved.

Digitisers are available in various designs and sizes. Those used for encoding map data must feature high resolution and accuracy. A common set of characteristics includes a resolution of 0.02 mm (the smallest unit of registration), an absolute accuracy of ± 0.15 mm and a repetitive accuracy of ± 0.025 mm.

Some digitisers are fitted with background illumination as are ordinary light tables, an advantage in carrying out many digitising tasks. Overall size ranges from the 27 X 27 cm tablet size up to 1.0 X 1.5 metres. Map digitising

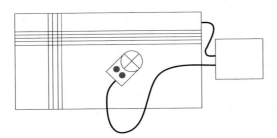

Fig. 5.7 – Most digitisers are basically a flat surface in which is embedded a grid of thin wires for conducting electric current. The sensor coil detects the electromagnetic field around the wires, thereby registering the position of the cross hair centre.

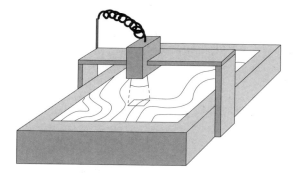

Fig. 5.8 – Scanners are operated by means of a photosensitive pickup being moved with great precision over the surface of a map. Photosensitive cells register the different grey tones on the map and thereby create a raster image showing all the lines and white areas on the map.

requires a relatively large surface capable of accepting the common 84.1 X 118.9 cm ISO A0 format (according to the international ISO 216-1975 standard).

Scanners

Scanning is a means of encoding through conversion to a digital image comprising a regular grid of pixels, rather like the array of dots that comprise a newspaper photograph. Scanners classify into three main categories according to their basic mechanism and physical size. These are flat-bed scanners, drum scanners and continuous-feed scanners.

Flat-bed scanners resemble automatic drafting machines. Map image information is registered via a photosensitive pickup that rides on a track, which is mounted on a beam that itself moves in the transverse direction. Scanners can be extremely accurate and capable of high resolution. Normal pixel size can be as small as 25 to 50 microns (0.025 to 0.050 mm), which corresponds to a resolution of 2.5 to 5.0 cm in detailing terrain from a map with a scale of 1:1,000.

Drum scanners employ the same process as flat-bed scanners, except that the map to be encoded is mounted on a drum which rotates beneath a fixed track on which the pickup rides. Normally the pickup digitises one grid line for each drum rotation.

With continuous-feed scanners, documents to be scanned are fed past a row of sensors as wide or wider than the maximum width of the documents scanned. The basic structure is simpler and less expensive than that of table or drum scanners, so accuracy and resolution are limited accordingly. Therefore, continuous-feed scanners are best suited to ordinary document scanning, which is why they often are called `document scanners'.

Scanners can register colours, usually in a three pass process involving trans-illumination of the scanned map and successive magneta, yellow and cyan blue filtering.

Some scanners convert raster data to vector data in the scanning process itself, and simultaneously present the results on screen.

Scanners have a drawback that sometimes must be overcome. Scanning senses everything on a map, including map labels, coffe stains, wrinkles, and other unwanted information. Therefore, extensive editing is almost always necessary unless a map is drawn as needed, which rarely is the case. As editing invariably is expensive, many organisations now routinely redraw maps to be scanned.

Video

Video cameras are also used to digitise maps, documents and pictures. The resolution and accuracy attainable are relatively poor (typically 512 X 512 pixels), although adequate for some applications.

Automatic Line Tracers

An automatic line tracer may be used rather than manual tracing on a digitiser whenever the data calls for individual lines to be scanned. The automation involves an "intelligent" scanner, a narrow-beam laser which follows lines. However, the operator must intervene and lead the scanner further whenever it encounters an intersection, the end of a line or another conflict.

Coordinates are computed continuously for all points on a line, as they are by a digitiser. Automatic line tracing is slower than raster scanning but more rapid than manual tablet digitising, and the results are lines in continuous form (vector data).

Software programs have been developed to perform an automatic line tracing function on data that has been scanned and displayed on screen.

5.1.4 PLOTTERS AND OTHER OUTPUT DEVICES
Pen Plotters

Pen plotters are output devices that produce continuous lines. The are sometimes are called vector plotters because they can only process and plot vector data. There are two types: flat-bed plotters and drum plotters. Larger plotters often contain built-in hardware and software, which simplifies drawing and improves their graphic capabilities.

Flat-bed plotters have electric motors to control drawing motions in the x and y directions, on a drawing medium fixed to a flat surface with tape or held by vacuum. One or more pens are mounted onto a carriage which moves along a beam in the y direction. The beam moves on tracks in the x direction. The carriage may hold ball-point pens, stylographs or felt-tip pens in various colours, while some plotters may have special carriages fitted with engraving scribes or photo heads. In this way, drawings may be made on paper, plastic, engraving foils or photographic paper.

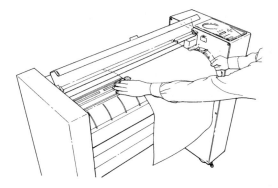

Fig. 5.9 – A drum plotter of Hewlett Packard type HP Draft Master.

Flat-bed plotters are relatively accurate, typically with an absolute position accuracy of ± 0.075 mm and a repetition accuracy of ± 0.015 mm in both axial directions. The largest plotters can handle formats up to ISO A0 (84.1 X 118.9 cm).

Although large flat-bed plotters are available, drum plotters are often preferable because they can handle still larger drawings and comparatively may be less expensive. The basic drum plotter mechanism for moving pens back and forth on a beam is similar to that of the flat-bed plotter. The beam, however, is fixed parallel to the axis of a drum, and the drawing devices are often placed in a carousel at one end of the beam. The drawing medium is wrapped partly around the drum, which rotates in both directions to provide motion transverse to the beam axis. The accuracy of a typical drum plotter is about ± 0.25 mm.

Smaller plotters, for ISO A3 (29.7 X 42.0 cm) and A4 (21.0 X 29.7 cm) formats, are often hybrids of the flat-bed and drum designs, in which the medium is fixed by a drum to a bed. Accuracy is usually in the order of ± 0.25 mm.

Electrographic Plotters

Electrographic plotters include electrostatic, eletro-photographic and electrosensitive plotters. The electrostatic and electrophotographic processes, which are used most frequently in GIS applications, involve an intermediate electrostatic image. The electrosentive process involves direct interaction between an electric current and the surface of a special paper, and is used in various high-resolution applications such as photo-typesetting.

In an electrostatic plotter, the image is first written as a pattern of electro-static charge, either directly onto specially-treated paper or onto a drum coated with aluminium oxide. The pattern is then made visible by washing the charge with a collodial suspension of particles of pigment carrying the opposite charge. In the drum design, the visible image is then transferred to plain paper by the application of pressure and an electric field.

Electrostatic plotters, which are frequently used in GIS applications, have

Fig. 5.10 – An electrostatic plotter, Calcomp Model 5835XP.

an advantage over pen plotters in that they can plot from both vector and raster data. The drawback, however, is that most models require special paper, which means that existing maps cannot be amended. The resolution of quality electrostatic plotters is usually in the order of 200 to 400 points per inch.

In an electrophotographic plotter, a beam of light writes the image onto the electrically charged surface of a drum. The light discharges the charged surface to form the image, which becomes visible once the drum is dusted with a dielectric powder, called a toner. The image is then transferred to paper by direct contact and the application of an electric field. The toner particles are finally affixed to the paper by heat and/or pressure.

The electrophotographic process, also known as xerography, is the same as that used in most office copying machines. It is also used in laser printers, in which lasers are the light sources. A wide range of laser printers is now available for use with text and graphics, including the printing of map data. Most handle paper formats up to standard letterhead: ISO A4 (21.0 X 29.7 cm) or the US 81/2 X 11 inch standard, but printers for ISO A3 (29.7 X 42.0 cm) are also available. Laser printers have resolutions varying from 300 to 600 dots per inch (dpi). Colour laser printers are now available with digital inputs able to receive GIS data and produce maps.

All electrographic plotters can produce final images in black-and-white or an array of colours, depending on the pigment colour (magneta, yellow, cyan blue and black, in successive passes).

Thermal Plotters

Thermal plotters and thermal transfer plotters use heat to produce images. Thermal plotters use local heating to warm thermo-sensitive paper, which is coated with two separate, colourless components. Once heated, these combine to produce a wide range of colours of fair quality. Thermal transfer plotters transfer coloured wax to produce high-quality images. Both types are available to handle sheet sizes up to ISO A3 (29.7 X 42.0 cm).

Ink Jet Printers

Ink jet printers form images by projecting droplets of various coloured ink onto paper. The technique has long been used in the printing and labelling industries and was first used in computer printout devices in the mid- 1970s. One variety, the pulsed ink jet printer, in which one or more columns of ink nozzles are mounted on a head that traverses the medium, is capable of high printing speeds, up to 400 characters per second. From a cartographic viewpoint, the accuracy of most ink jet printers is relatively poor. Therefore, they are mainly used for such low-accuracy applications as replicating screen displays. However, in the early 1990s ink jet printers with high resolution, up to 400 dots per inch, became available, both in letter sizes and in larger sizes suitable for use as a plotter substitute.

Film Printers

In a film printer, a laser draws a raster image on a photosensitive medium. The image resolution is fine, up to 300 lines per centimetre, the accuracy good. The technique is used to produce film originals for printing, as used for printing maps in colour.

Digital map data may also be directly entered into devices that produce films for printing.

5.2   SOFTWARE
5.2.1   GENERAL

Geographic information systems often comprise complex program systems for various applications. This is because many systems were originally devised for special, limited tasks, such as mapping the network of a particular energy supplier. However, the current trend is towards more general systems, or "tool kits" from which users may select components suited to their own specific applications. The tasks involved can be demanding, so general systems often have relatively high entry levels.

Many software systems are constrained to use with specific computers and operating systems. There are also constraints on the types of graphic displays, plotters, digitisers and other peripheral devices which may be used. Moving a software system from one computer to another can be extremely difficult because of the difference in the respective operating systems. However, the process is becoming easier now that standards have become established in the market.

Modern software systems are flexibile and to a great degree based on official or de facto standards. One such standard is the Graphics Kernel System (GKS). This set of computer graphics routines is intended for use by application programmers, and is internationally standardised by the International Organisation for Standardisation (in the ISO 7942 and ISO 8651 standards). GKS assures consistency. Thus a screen can be identified as a screen regardless of its maker. De facto industry standards have also been established. An example is HPGL, a plotter language developed by Hewlett-Packard, and PostScript, a page description language developed by Adobe Systems for word processors and desktop publishing packages.

Many software systems boast libraries containing a wide range of types and models of digitisers, screens, plotters and other peripherals, which the software supports through having the relevant drivers. A driver is a software routine that handles the specific details and characteristics of a single peripheral device so that it can work correctly with the system. The user may select peripherals from the library listings in order to customise the software to a particular set of peripheral devices.

5.2.2   COMMUNICATIONS BETWEEN USERS AND COMPUTERS

Users and computers are said to communicate at the Man-Machine Interface (MMI). The features of MMI are vital in GIS applications, where users need to control communications with their computers by deciding what is to be done at what time and in what order – always with the option of cancelling an action midway through a communication.

In particular, MMI communications require input/output devices, such as ordinary keyboards, special function keys, mice, joy sticks, etc., along with their supporting software. More recently, tablets have been used in interactive man-machine communications. A tablet is a graphical input device in which positions can be selected by using cross-hairs instead of pointing on screen.

COMMANDS (ArcInfo, ESRI)

– RESEL road ARCS route > 0 AND veh CN 'T'
   (logical selection among road and vehicles)

– Lineset last.lin
   (activating symbol table)

– ARCLINES road route route.lut
   (ploting instructions for selected roads using
    route no. to select symbol through look up-table)

Fig. 5.11 – System commands can be given in the form of long strings of command instructions.

PROMTS

Give no. for selected route of roads	: <	>
Select roads maintained by lorry (L) or truck (T)	: <	>
Plot the selected roads, Y / N	: <	>

Fig. 5.12 – Command activation can be in the form of questions to the operator. These questions can often be adapted by the user for special functions.

The dialogue between a user and a computer is said to be interactive when user commands result in immediate system responses in the form of messages, error messages, new choices, etc.

On-screen dialogue may be controlled by:

- user commands to the system
- user responses to system messages or queries
- menus of options that elicit user input
- icons (pictorial symbols used in menus) upon which users point the cursor

Commands

The first software systems were controlled by relatively cumbersome sets of commands and had no interactive features. Many GIS contain remnants of these earlier complexities, although choices between commands and menus are often available. To use commands, users must enter relatively long instructions which in turn require a knowledge of numerous basic commands.

Modern systems are often structured so that users may compose and store sequences of composite commands compiled from large registers of basic commands. The programming involves system macro instructions. Composite commands are best suited to repetitive operations for a specific product or application.

Tone Table Picture	Standard Table Picture	Align Edges	Align Corners	Align Centers	Align Reference Points		
	Get Mis	Create Mis	Undelete Mp	Undelete map	Current Mp	Current Visible Picture	
		Delete Mp	Delete Visible Picture	Position Picture	Move Any Picture	Move Visible Picture	List Mp

Fig. 5.13 – Plotter menu from the Swedish Teragon image-treatment system. Commands are activated by positioning the menu table on the digitiser and pointing and clicking the mouse in the appropriate fields.

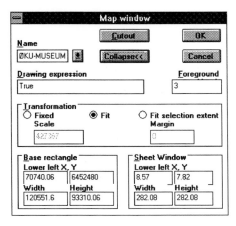

Fig. 5.14 – In modern GIS, dialogue with the user is generally based on screen menus where one can point and click in the command fields. (MapView© UVE)

Queries and Messages

The dialogue may involve the user either answering system queries or following instructions given in system messages. Examples are given in Fig. 5.12.

Dialogues of the type illustrated in Fig. 5.13 are well suited to the needs of many end-users, for instance those who use PCs or workstations in serving the public.

Menues

Communication between users and computers via menus is now commonplace. A menu is a program that elicits user input through an on-screen display of options, which may be commands or keys to sub menus. The software system is then said to be menu-driven.

An option may be selected from a menu by moving the cursor to it, using the arrow keys on the keyboard, and striking a single key. Or, as is now more common in GIS applications, a mouse or other pointing device may be used to position the cursor on an option, which is then selected by clicking the mouse button or another key.

On-screen menus are commonly organised with a row of the titles of sub-menus across the top of the screen. The sub-menus are called pull-down menus, because when selected they appear to be pulled down from the top row, as a roller shade is pulled down from the top of a window frame. Options in the pull-down menus may, in turn, trigger more pop-up menus. Functions such as these are available in a variety of "windows" programs (including MS-Windows, X-Windows, etc.)

Although though most software systems are now menu driven, many experienced operators prefer commands to menus because commands permit greater flexibility and often achieve final results more rapidly.

Icons

Icons are selected by pointing-and-clicking. Each pictorial icon on screen represents a command. The icons may be simple, such as a set of boxes for colour selection. Or they may be more complex, such as a rubbish bin to represent halting and discarding.

Fig. 5.15 – Command fields in screen menus can be presented in the form of easily comprehensible symbols called "icons".

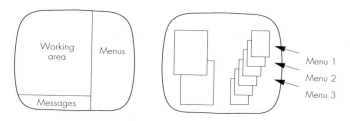

Fig. 5.16 – The screen can be split into several fields by means of the "windows" technique. This allows several different commands to be activated simultaneously.

Windows

Windows techniques permit simultaneous operation of differing programs by displaying information on them in separate rectangular areas on screen.

Windows, menus and dialogue boxes may be moved around on screen as needed, for example, when they obscure part of an active application on screen.

Today, there is a growing trend towards the use of simple and economical PC-based information and presentation systems as a supplement to the heavier GIS software. These new systems are often based on windows or windows-equivalent Man-Machine Interface.

Ancillary Feautures

Working screen space is often scarce, particularly if it is blocked by menus or other software tools or if both the details of and an overview of an operation must be viewed simultaneously. Therefore, some systems use two screens, one as a primary working screen and the other for communications and enlargements. The screens, a digitiser and a tablet are then usually combined in a work station.

Many systems contain help functions which users may activate for further information on commands. Once a command is activated, an explanation of it can be requested using a contextual search.

Users often may select various function parameters. If a parameter is not specified by the user, the system will revert to its own standard value, known as the default value.

Response time is crucial in all interactive work, as response times longer than a tenth of a second cause a user continuously to doubt that the system has understood the last command entered. However, in some cases, as with final drawing, batch processing may be advantageous. In batch processing, commands are collected in a file and executed when computer resources become available.

Some systems can accept user program modules in addition to the interactive menu-driven system.

5.2.3   USER REQUIREMENTS

GIS users may be divided into groups according to their functions and needs. These groups might comprise:

- end users, who seek problem solutions and only see final products in the form of maps and reports
- GIS specialists, who advise and solve problems for end users, GIS operators and data compilers
- GIS operators, who understand the functions of a specific system so as to manipulate the data
- data compilers, who understand the data but not the system

The value of GIS depends largely upon how well users state the problems and how well GIS specialists provide solutions. End users frequently need only a limited knowledge of GIS. However, GIS specialists must be familiar with the general principles of GIS, data and data compilation, cartographic facilities and, not least with the potentials and limitations of the system employed. In other words, a specialist must be able to sketch alternative production routes.

A GIS specialist must be expert in several fields, though not all in a GIS organisation need the same capabilities. Outside consultants may also be engaged.

Normally, it takes an operator about a year to become thoroughly familiar with a larger system. An operator can, however, can be productive long before that. In many cases, it is possible to complete without a specific knowledge of all the available functons.

The initial training of operators preparing to work on larger systems usually lasts 14 days, and is followed by additional training for specific parts of the system. Furthermore, whenever new computers are installed, training courses are held to familiarise staff with the operating system, backup routines and other facets of their operation.

The need for computer knowledge depends on the size of the system and the task confronting the operator. A single-user PC is easier to operate than a multi-user minicomputer system. Professional staff members need only moderate computer knowledge in order to learn how a system works. That said, extensive computer knowledge is a prerequisite for staff who will be responsible for operating the system and will require an in-depth understanding of the geographic information system, the operative system and the peripheral equipment.

Most organisations that have implemented GIS so far have been obliged to conduct some development for matching older data systems (such as those containing older registers), improving existing programs, interfacing new peripheral equipment, and so on. Developments such as these require computer expertise which may be acquired in-house or contracted externally.

Digitising personnel and other data compilers should be familiar with the attributes with which they deal, but need only a limited knowledge of the GIS involved.

5.3 WORKING ENVIRONMENT

Work at a GIS station can be demanding. Therefore the working environment involved in an GIS implementation should always be taken into consideration.

Digitising maps can be physically tiring and mentally tedious. Clearly, such work cannot be performed continuously for an entire working day, and should be split up through rotation or with longer breaks. Spring-loaded armrests are available to ease arm loading in digitising. The same precautions apply to the prolonged use of a mouse for pointing and clicking.

A good working position and sufficient breaks are also necessary for good work at a screen.

Work places should be relatively compact, so that manuals, maps, manuscripts, notes and the like are all within reach. A U- or L-shaped layout is often best. Screen height above the floor and distance from the user should both be adjustable. Keyboard and tablet positions should also be freely adjustable. Illumination should be indirect to minimise screen reflections.

As concentration is vital in GIS work, work places should be shielded against disturbances. Printers, plotters and other noisy equipment should be located away from the operator. In some cases where equipment produces heat, air conditioning or fans may be required.

Staff training is essential to the success of the system, and system handbooks and other documentation should preferably be in the operators' mother language.

6. SOURCES, COLLECTION AND INTEGRATION OF DATA

6.1   INTRODUCTION

The data sources for a comprehensive GIS are probably both more numerous and of greater variety than in most other information systems. Indeed, GIS could rightfully be called a "mixed-data system." Most often, the data mixture makes the system lively.

At the outset, data used in GIS may be:
- In various digital forms – vector, raster, various databases, spread sheet tables, satellite data, and so on.
- Non-digital graphics, such as conventional maps, photographs, sketches, schematic diagrams and the like.
- Conventional documents in registers and files.
- Compilations in scientific reports.
- Collections of survey measurements expressed in coordinates or other units.

Although most of the problems associated with collecting data for GIS uses had been identified by the mid 1970s, data collection remains the most expen-

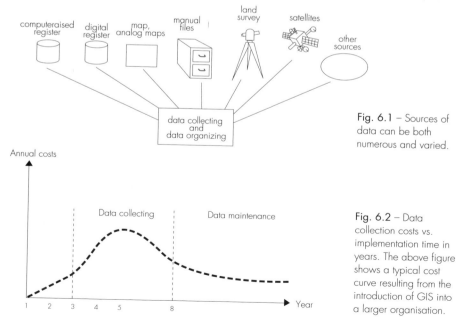

Fig. 6.1 – Sources of data can be both numerous and varied.

Fig. 6.2 – Data collection costs vs. implementation time in years. The above figure shows a typical cost curve resulting from the introduction of GIS into a larger organisation.

sive and time consuming aspect of setting up a major GIS facility. Experience indicates that data collection accounts for 60% to 80% of the total cost (time and money) of a fully operational GIS, whilst the purchase of equipment accounts for not more than 10% to 30%. Other costs include training, development, administration, and so on. As shown in Fig. 6.2, the costs of data collection are dominant even when spread out over an implementation period of several years.

6.2   DIGITISING MAPS

There are five methods of producing digital map data:

1. Digitising existing maps with a digitiser.
2. Scanning existing maps.
3. Digital photogrammetric map production.
4. Manual entry of measured and computed coordinates.
5. Transfer from existing digital sources.

The first three methods, which produce digital data from ordinary maps, may all be divided into three phases:

1. Preparation
2. Digitisation/scanning
3. Editing data

6.2.1   MANUAL DIGITISING

Preparation
Data must frequently be prepared and carefully organised in order to facilitate and expedite the digitising work itself.

Manuscript maps may be used to group and highlight information relevant to the various themes involved. For instance, roads shown on a base map may simply need highlighting in various colours, according to the type of road. Or the operation may be more comprehensive, as when collecting and grouping network information from various maps and sketches onto a single common map. In practice, these operations invariably demand some redrawing of maps prior to digitisation. As a rule, changes and amendments to graphics are achieved more easily on paper maps prior to digitisation than they are on digital data after digitisation. Experience has shown (S. Aronoff/ U.S. Forest Service 1991) that even when manuscript maps must be redrawn extensively, they can save as much as 50% of digitisation costs. Consequently, manuscript maps are considered vital in digitisation.

The identifier data requisite to relating digital map data to computerised attribute data are usually organised on manuscript maps.

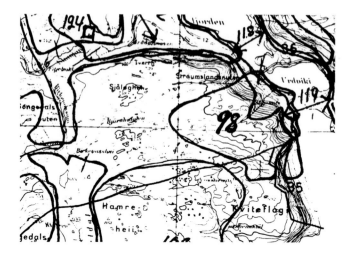

Fig. 6.3 – A typical manuscript map showing game registration in Norway.

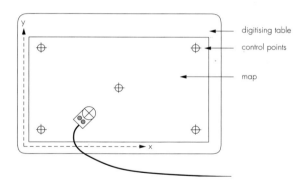

Fig. 6.4 – Map digitising. A digitising table complete with map and with the table's internal coordinate system defined.

digitising table

control points

map

Digitising

As illustrated in Fig. 6.4, digitisers are used in manual digitising. Digitisers have their own internal coordinate systems which may be related to terrain coordinates by cross-registering three to six or more coordinates with known terrain coordinates, such as grid intersections, polygon points or border markers. These cross-registrations permit the computation of transformation parameters, which may then be used to compute terrain coordinates from digitiser coordinates. Some programs convert coordinates continuously during digitisation, whilst others convert in the editing phase.

Conformal transformations involve transferring digitiser coordinates to terrain coordinates through parallel translation, rotation and scale alteration. All objects then retain their original shapes. Affine transformations alter the shapes of objects, through unequal x and y scale alterations and unequal rotations of the x and y axes. The mathematical relationships are:

Conformal transformation:

$$X = c_1 + ax - by \qquad\qquad (6\text{-}1A)$$
$$Y = c_2 + ay + bx \qquad\qquad (6\text{-}1B)$$

Affine transformation:

$$X = c_1 + a_1x + b_1y \qquad\qquad (6\text{-}2A)$$
$$Y = c_2 + a_2x + b_2y \qquad\qquad (6\text{-}2B)$$

where X and Y are terrain coordinates and x and y are digitiser coordinates; a, b, c_1 and c_2 are conformal transformation parameters; and a_1, a_2, b_1, b_2, c_1 and c_2 are affine transformation parameters.

Once the known points are registered, digitising may begin. Two methods are used. In point digitising, a cross-hair or cursor is placed over the position to be recorded and a key is pressed to enter the datum. In stream digitising, a cross-hair or cursor is drawn along a linear symbol and data on positions are automatically entered at pre-set intervals of either time (such as 5 per second) or distance (such as 5 per centimetre). Data entry may also be curve fitted, with more entries as the curvature radius decreases.

Point digitising requires that the operator continuously selects points and keys in data entries. Therefore it is most suitable for individual points and for lines comprising straight segments, such as property boundaries and building outlines, which need not be followed between successive data point entries.

Stream digitising requires more sophisticated computer hardware and software (partly because data entry speeds may be high) and greater operator skill. Once these requirements are met, stream digitising is well suited to digitising sinuous lines, such as contours and the borders between different soil types. The drawback is that unless its parameters are selected carefully, stream digitising may generate unnecessarily large quantities of data.

Point digitising is preferable to stream digitising for high accuracy entry.

Lines of fixed curvature, such as property borders to road curves, may be described by radii of curvature rather than by their constituent points.

Fig. 6.5 – Principle of stream digitising.
Different methods can be applied in digitising to select a point on the line, based on, among other factors, a given arrow height value as shown in the illustration.

Fig. 6.6 – Digitising errors. Typical errors which can occur during digitising. There are three types of error which often occur during digitising. Such errors can result in considerable correction work. The type of error illustrated in fig. b) where dangles occur can often be most easily corrected automatically. It is therefore generally recommended that, during digitising, situation b) is preferable to a). The error in fig. c) creates a new meaningless polygon.

Digitising includes the entry of thematic codes for object types and i.d. codes which link the object types to attribute data. For instance, digitising a building shown includes entry of the thematic code for buildings and an i.d. number for the specific type of building. A new i.d. number is entered for the next building, and so on. Codes are keyed in or entered via a tablet menu.

In some systems, the i.d. codes of map objects may be entered directly, provided stored attribute data is available interactively on screen.

Digitising accuracy and progress should be verified as work proceeds. Verification may be manual or semi-automatic. The former requires objects on a manuscript map to be crossed off as they are digitised and for partially completed maps to be printed out as the work progresses. In some systems, results may be continuously displayed on screen, which in effect is a semi-automatic, man-machine variety of echo check (a comparison of received/ entered data with original data). The screen display enables the operator to verify the work performed and to avoid double entries.

Digitising invariably introduces error. Typical errors include contour lines that cross each other; gaps between lines that should be contiguous but dangle, or fail to meet; intersecting lines which should not overlay yet do; lines meeting at the wrong point or intersecting each other; objects with improper or missing codes; and missing details.

Some errors are corrected more easily than others. For instance, it is better that the digitisation of two lines contiguous at a point result in their crossing rather than in their not meeting at all. Crossed lines are more easily processed, particularly as the original contiguous lines (and their incorrect crossings) are normally parts of the border of a closed polygon, and a polygon of incorrect geometry is more easily corrected than a polygon that remains open.

Some systems support snapping on, a program for recognising and con-necting to nearby points or lines, so that nodes, closed polygons and con-tiguous lines may be formed quickly.

Skilled operators can achieve cross-hair placement accuracies of ± 0.05 mm to ± 0.06 mm (P.v. Bolstad, P. Gessler, T.M. Lillesand 1990).

The time required for digitising depends on the complexity of the map in question and the extent of coding. Digitisation of a map showing 300 soil types can involve eight hours of data entry and four hours of subsequent editing (Burrough 1986). A rule of thumb holds that digitising may progress at a speed of one polygon per minute (NCGIA, Core Curriculum 1991).

In most cases, digitising may be performed using a program run on a PC. This is both convenient and quick, firstly because there is no need for other computing facilities and secondly because PCs now have the capacity and speed to suit the task. Error correction, editing, topology structuring and other tasks may also be performed using suitable programs.

Operator skill is decisive, both in terms of the time required for digitising and for its quality. Most GIS facilities have found that operators experienced in cartography are preferable to operators experienced in computers or other disciplines. Operators must be trained to understand the logical connections between property borders, pipe networks, soil type borders and other themes that are digitised, and to master the software tools they use to the extent that they can correct their own digitising errors.

Editing Data

Normally, the product of manual digitisation must be edited. Tasks include:

- error correction
- entering missing data
- forming topology

Data are verified visually on screen and/or on test printouts. Geometry, i.d. codes and any thematic codes may be plotted and verified. A test printout may be laid on top of an original on a light table and compared to it, bit by bit. Whenever many themes are digitised, this comparison is made easier if each theme is printed out and compared to the original separately.

Errors may be corrected and missing data may be entered either in the digitising program or in the GIS programs. Editing data, however, is often time consuming. When the original data is poor, editing may take more time than digitising itself. A program can, for instance, identify errors of the type illustrated in Fig. 6.6 and automatically delete unnecessary line segments, connect lines and close polygons. However, automatic programs are no panacea. In practice, they often correct dangles and gaps which, in fact, should be left alone.

Forming topology also reveals many errors. Typically, the same polygon number will be assigned to both sides of an unconnected line in a polygon structure. This is illogical, as all lines in a polygon structure must mark the borders between adjoining polygons.

6.2.2   SCANNING

Maps are scanned in order to:
1. use digital image data as bases for other (vector) map information
2. convert scanned data to vector data for use in vector GIS

Preparation

Scanning requires that a map scanned:
- be of high cartographic quality with clearly defined lines, text and symbols
- be clean without extraneous stains
- have lines of 0.1 mm and greater widths

In scanning tasks, work is often moved from the preparatory phase to the editing phase. If, for instance, the data encoded are to be used directly as digital image data, the quality requirements for the scanned map may be relaxed and preparation may be negligible.

Scanning

Scanning comprises two operations, scanning and binary encoding. Scanning a map produces a regular grid (pixels) of grey scale levels. Grey scales often range numerically from 0 for complete black to 255 for pure white. Pixel sizes range from 25 µm to 50 µm. [µm is the symbol for micrometer, one millionth of a meter, or one thousandth of a millimetre. Its former name, "micron" and former symbol, "µ" were dropped on 13 October 1967 by action of the 13th General Conference on Weights and Measures. The symbol "µ" is now used exclusively to designate the prefix "micro-," which means "10^{-6}" or "one-millionth."]

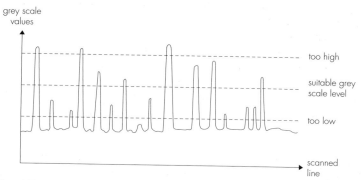

Fig. 6.7 – Grey scale level. This figure shows the grey scale values along a scanned line. The peaks appear as lines on the map, whereas the bases are shown as white areas. Tests are carried out to establish the correct grey scale value for each map being scanned.

55	56	81	57	51	49	44		0	0	1	0	0	0	0
47	53	89	60	50	59	48		0	0	1	0	0	0	0
56	63	81	52	47	45	42		0	0	1	0	0	0	0
54	60	63	88	56	50	46		0	0	0	1	0	0	0
59	66	66	70	89	67	54		0	0	0	0	1	0	0
46	57	60	71	81	65	51		0	0	0	0	1	0	0
45	55	57	66	88	64	59		0	0	0	0	1	0	0
50	51	60	56	69	85	67		0	0	0	0	0	1	0
51	53	59	64	66	70	80		0	0	0	0	0	0	1

Fig. 6.8 – Digital raster image – grey level array/binary. On the left of the illustration is a raster image with grey scale values. The corresponding binary encoded raster image is shown on the right. All grey scale values equal to 80 or more are lines on the map and have been given binary representation 1, whilst the remaining grey scale values are white areas on the map and have been given binary representation 0.

As a black/white line map comprises only black lines against a white background, scanning must clearly separate black from white. Accordingly, the scanner must be adjusted precisely for a grey scale level corresponding to the map demarcation between black and white. If the level is set too high, vital information represented by thinner lines, may be lost. If it is set too low, extraneous information, such as dirt, stains and dust, may be included.

Finding the most suitable grey scale level may be time consuming, but once it is found, a map may be scanned in minutes.

With a threshold set to differentiate black from white, a grey level array may be binary encoded. As shown in Fig. 6.8, each pixel is then binary coded as one or zero, depending on whether the grey level is over or under the threshold set.

Scanning produces voluminous data. With a pixel size of 50 μm, there are 400 pixels per square millimetre and six pixels in a line 0.3 mm wide. A 60 X 80 cm map encodes to about 200 million pixels.

Pixels are stored in rows and columns, and many adjacent pixels have identical values. Therefore, programs are available to compact the aggregate amount of data stored. Binary data is particularly suited for compression: run-length encoding compresses data by about 90% (G.D. Foley, A. van Dam 1990), and some advanced compression techniques can attain 98% (Robb, 1986).

Usually, a collection of data have a prefix header, which contains such details on the data as the transformation parameters, terrain coordinates of the lower left and upper right corners, the pixel size, and so on.

Editing Data

The editing phase of scanned data may include:
- upgrading raster data through division into three categories: shapes, potential symbols and noise

(• pattern recognition of symbol candidates in raster data)
- thinning and vectorisation
- separating vectorised data into shapes, potential symbols and noise
- vector data pattern recognition of symbol candidates
- error correction
- coding
- supplementing missing data
- forming topology

"Noise", a term borrowed from acoustics, designates such information as stains and extraneous lines that are registered but not required. Data may be improved by structuring binary information in three categories; shapes, potential symbols and noise. The noise category can then be filtered out, together with the potential symbols.

If the data are to be used only to compile a digital map image, only the noise need be filtered out.

Raster data contain no codes for map objects. But whenever map information is divided amongst various overlays (topography, inventories of facilities, land use, etc.), each overlay may be scanned individually to produce data relevant to its single theme. Each thematic overlay can then be converted to vector data with a common thematic code.

Raster to Vector – Vectorisation

Some scanners vectorise as they scan. Otherwise, vectorisation is usually part of the subsequent editing process. The vectorisation of structures (lines, symbols, etc.) may be summarised in six steps:

1. A number of pixels forming a structure, such as a line, are registered.
2. All pixels transverse to the line, save those in its middle, are stripped off (skeleton plotting).
3. Starting at one end, the pixels are connected, one by one along the line (linearisation).
4. Line curvatures are checked against set maxima, which if exceeded, indicate that the line is no longer straight, so linerisation terminates.
5. Coordinates are determined for the start and end points of the straight line segment, and a vector along the segment is formed accordingly.
6. Little by little, lines and structures are assigned coordinates and vectorisation continues.

The result of vectorisation is a continuous structure delineated by a sequence of coordinates designating the mid-line of the structure. A magnified illustration of a line of pixels is shown in Fig. 6.9.

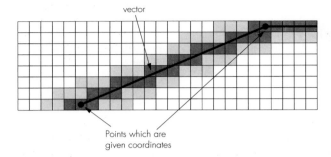

vector

Points which are
given coordinates

Fig. 6.9 – Vectorisation image. Vectorisation of a scanned line on the map. The darkest pixels are the skeleton which is used under vectorisation. The light rastered pixels have been stripped off, whilst the pure white pixels appear as a white area on the map.

Vectorisation may be subject to errors and defects, including:
- deformations or interruptions of lines intersecting at nodes
- vectorisation of extraneous stains and particles on the original map
- vectorisation of alphanumeric information and text
- unintentional line breaks resulting in divided vectors
- dotted line symbols (trails, soil type boundaries, etc.) resulting in many small vectors
- smooth curves become jagged, i.e. introduction of unwanted inflection points

Many systems include error correction routines in order to circumvent these problems and assure scanning quality.

Some systems have a tracer tool, which, when started in a line structure, follows it more or less automatically on screen. The process includes continuous vectorisation of raster data. All operations are on screen, so the operator needs to intervene only in conflicts. Thus, pipe networks may be scanned with only pipe routes, and no other map information, being vectorised.

Pattern Recognition

In scanned data, numbers, letters, symbols and other map information cannot be differentiated from one another directly; all information comprises either pixels or grey levels. However, pattern recognition programs are available to analyse raster and vector data, and to elicit features of a known pattern. The recognition of shapes and symbols is best performed on raster data because as vectorisation invariably involves a loss of information. Nonetheless, raster data are usually vectorised prior to recognition.

To ease recognition, vectorised data are usually divided into data concerning shapes, and relevant to potential symbols (after noise data has been filtered out).

Recognition of letters and numbers is particularly useful, as is that of lines and specific map symbols.

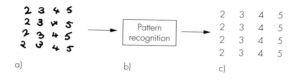

Fig. 6.10 – Pattern recognition. The figures in the binary raster image (a) are thinned, vectorised and pattern-recognised (b) and then translated to digital numerical values (c).

Pattern recognition usually involves a system that is "trained" to detect specified patterns by assigning probabilities to their possible presence. Patterns are usually specified analytically or with the use of templates (model patterns) for comparison.

For example, pattern recognition is used in the scanning of nautical charts in which the original registrations often comprise innumerable depth numbers and letters. Recognitions as high as 95% to 98% have been attained (K. Bråthen, E. Holbæk-Hansen, T. Taxt 1987). However, results depend strongly on the quality of the original.

Coding and other Editing og Vectorised Raster Data

Scanning and vectorisation invariably introduce error and obliterate some data. Consequently, data must be verified and corrected. Vectorised data have no thematic codes, so the various map themes cannot be distinguished (unless pattern recognition has been employed) and map objects have no i.d. codes that link them to attributes. Therefore, the data must be coded by assigning codes to individual objects. This is usually achieved manually on screen. Once vectorised data have been coded, the subsequent manipulations are the same as for vector data.

Integrating Image Data in GIS

Some GIS can manipulate and display both vector data and raster (image) data on screen. In principle, raster data may be:

- scanned maps
- scanned documents, photographs, etc.
- satellite data
- video images of maps, photographs, aerial photographs, etc.

Vector data and raster data may be displayed simultaneously on a screen supported by storage split between raster and vector data, and accompanied by an editing program for both types of data.

In order to be meaningful, the two sets of data entered must employ the same coordinate system for the displayed images. The relative positions of pixels, stated in row and column numbers, are transformed to terrain co-ordinates using selected points whose coordinates are known, as with the

corners of houses, grid intersections and road intersections. The transformations may be conformal, affine or of another type depending on the
systematic errors in the scanned data.

Packed, stored raster data must be unpacked prior to being displayed on
screen. Again, unpacking may be time consuming, but once it is completed
and the unpacked data are stored in the working memory, the manipulation
response time is satisfactory.

Scanned map data may be used to compile a base image whenever the
map generated is for orientation only. An example is the pipe data of a
network to be documented in raster form, but for which the pipe routes are in
vector form. Other examples include satellite data and images integrated in
vector GIS. Applications of raster images include forming the bases for
interactive correction and the amendment of vector data.

The chief advantage of scanning maps to raster images is that:
• compilation of digital map images is rapid

The disadvantages are that:
• raster data files are large and consume memory space
• raster data lack "intelligence" and hence can be used only as bases for
  drawings
• the lines on enlarged raster maps may be annoyingly thick.

Scanned documents, video images and other inputs may be integrated in
GIS. Raster data may be stored on a video disk or a hard disk. A raster image
and the GIS into which it is entered are linked via a table of identifiers of the
map objects, and pointers that indicate their relevant storage locations.

Video images from a video disk can usually be screened more rapidly than
scanned map data from a hard disk, which must be unpacked from storage.

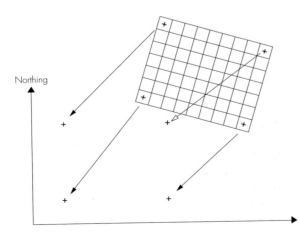

Northing

Easting

Fig. 6.11 – Transformation of
raster registrations to a
coordinate system.

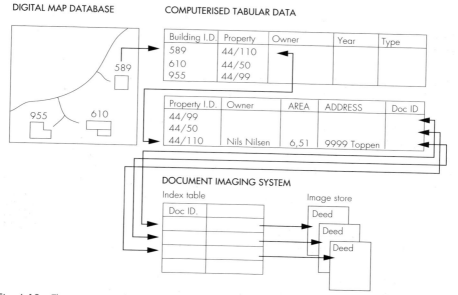

Fig. 6.12 – The integration of document and scanned images within GIS. For example, it is possible to point at a building and bring up an image of the corresponding deed.

6.2.3   MANUAL DIGITISING OR SCANNING

The advantages of manual digitising are that it:
- can be performed on inexpensive equipment
- requires little training
- does not need particularly high map quality

The disadvantages are that it is:
- tedious
- time consuming

The advantages of scanning are that it is:
- easily performed
- rapid

The disadvantages are that it:
- requires expensive equipment
- involves expert personnel
- usually entails considerable editing
- needs clean maps with well defined lines
- produces large quantities of data
- requires expensive raster plotters

Although scanning produces large quantities of data, compressed raster data need not require more memory space than vector data of customary format, i.e. ASCII. One reason is that raster data lacks the voluminous drawing instructions that are essential in vector data.

During the early 1990s, manual digitising has been (early 1990s) the technique used most frequently. But scanning may have the greater potential, primarily because in many applications speed is crucial. Further developments in pattern recognition will undoubtedly contribute to more widespread use of scanning.

6.3   SURVEYING AND MANUAL COORDINATE ENTRY

6.3.1   SURVEYING

In surveying, measured angles at, and distances from, known points are used to determine the positions of other points.

Surveying field data are almost always set out in polar coordinates, which may be transformed to rectangular coordinates using the relationships:

$$\Delta x = D\cos Gb \qquad \Delta y = D\sin Gb \qquad (6\text{-}3A)$$
$$X_p = \Delta x + X_1 \qquad Y_p = \Delta y + Y_1 \qquad (6\text{-}3B)$$

Heights are computed from horizontal or inclined distances. Vertical angles are computed between known points to new points.

$$\Delta e = D\cos Z \qquad (6\text{-}4A)$$
$$E_p = \Delta e + E_1 + i - s \qquad (6\text{-}4B)$$

Surveying has traditionally produced numerical data in the form of angles measured by transits and theodolites, and distances measured by tapes and

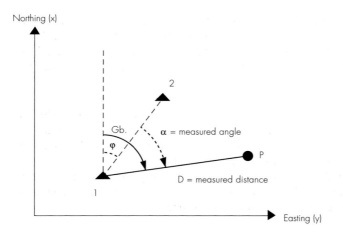

Fig. 6.13 – The geometric principles of surveying (polar coordinates). Polar coordinates with direction (Rv.) and distance (D) can be converted to right-angle coordinates.

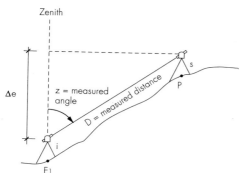

Zenith

Δe

z = measured angle

D = measured distance

i

E₁

S

P

Fig. 6.14 – Surveying measurement of height. Height differences can be calculated from the measured zenith distance and the given / measured distance.

Fig. 6.15 – These days surveying is generally carried out using modern "total station" equipment, known as total stations. With this equipment it is possible to carry out measurements in the field and register observations through an interface to larger computer equipment in the office. (Geodimeter).

chains. Starting in the 1950s, however, the traditional optical and mechanical tools of the surveying trade were gradually supplanted by more accurate electronic and electro-optical devices such as lasers for measuring distances. Modern surveying is computerised: a "total station" theodolite, as illustrated in Fig. 6.15, can now store and process digital measurement data either in the field or, subsequently, through an interface to a computer. Data may be coded or assigned themes in subsequent computer editing.

Surveying point location accuracy usually varies from ± 1 cm to ± 10 cm, depending how measurements are made. Measurement speed depends on topographical variations and the degree of modelling desired. As a guideline, surveying of complete terrain details can cover about one to two decares per hour.

Surveying is usually employed whenever current information is needed on property lines, buildings, manhole locations and the like, as well as for smaller areas and for details that must be measured accurately. Photogrammetry is usually employed in mapping larger areas.

Computed coordinates are most often used to supplement and update existing maps, and coordinate lists are compiled for data on survey control stations, properties, manholes, and other fixed objects.

6.3.2   MANUAL ENTRY OF COORDINATES

Survey data may be entered in GIS whenever accurate map data are required, for instance when determining property borders for legal purposes. Normally, coordinates are keyed in manually, either directly to GIS or via a simple registration program. Data for each property should be verified, by identifying nodes and closing polygons, as entry progresses.

This process is time consuming – two or three times longer than the corresponding task using a digitiser, inclusive of all preparatory and editing tasks. However, the data produced are more applicable than digitiser data.

6.4   AERIAL PHOTOGRAPHS AND PHOTO INTERPRETATION

The concept of aerial photography was patented in 1855 by French writer and photographer Gaspard-Felix Tournachon (1820 – 1910), who wrote under the pen name of Nadar. In 1856, Tournachon daringly ascended in a balloon from a village near Paris and took the world's first aerial photograph. In 1909, airplanes were first used. Since then, aerial photography has become a discipline with its own theory and practice. Because aerial photographs are taken from above, all objects in them appear as viewed from various vertical angles. So aerial photographs almost always require interpretation in order to provide information for mapping or depicting details as seen from the ground. Of course, almost all photographs used for their information content are interpreted in some way, but the interpretation of aerial photographs for mapping purposes is most common.

The combination of aerial photography and photographic interpretation provides information on relatively large areas without survey inspection on the ground. Roads, lakes and water, buildings, farmland and forests are clearly visible in aerial photographs. Other characteristics like various types of vegetation, soil and geological formations are more difficult to interpret. Accurate interpretation depends both on experience and verification by ground control. Consequently, skilled photographic interpretation has become a profession.

Interpretation is based on extracting information from the shape of objects as well as size, pattern, shadows, grey tone or colour, and texture (by comparison with larger contiguous areas). Indirect indicators, such as surface shape, run-off patterns, locations and seasons may also be used.

Aerial photographs may be black and white on film sensitive to wavelengths of 0.4 µm to 0.7 µm, colour on film with layers sensitive to the 0.4 µm – 0.5 µm (blue), 0.5 µm to 0.6 µm (green) and 0.6 µm to 0.7 µm (red) ranges, or infra-red on film sensitive to the 0.7 µm to 1.1 µm range. Black and white film is the most common. Infra-red film, however, may assist interpretation, parti-

Fig. 6.16 – Aerial photographs have a resolution of 0.02 – 0.03 mm and contain, therefore, a considerable volume of information. Panchromatic black and white photos are often used, as shown in the illustration. Information from the photo can be interpreted by means of grey tones, shadows, patterns, textures, etc.

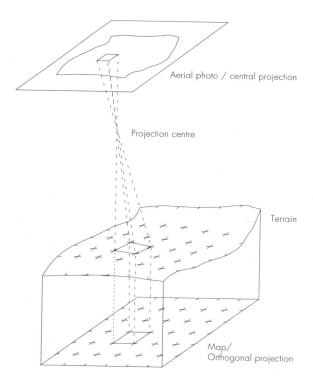

Aerial photo / central projection

Projection centre

Terrain

Map/
Orthogonal projection

Fig. 6.17 – Central projection and orthogonal projection of the terrain.

cularly of vegetation because plants emit varying radiation patterns through-out the infra-red spectrum. Most aerial photographs have a resolution, or smallest discernable detail, of about 0.025 mm.

Experienced photographic interpreters can map the details of vegetation, geology and other terrain features from aerial photographs and a few ground controls.
Almost all natural resource mapping is based on aerial photographs and photographic interpretation, as well as topographic maps.
Borders and characteristics of the various objects depicted in a photograph are drawn both in the field and in the survey office.

Aerial photographs, like most photographs, are taken through a lens that views a relatively large area and replicates it, inverted, in a smaller area on a negative. The geometry involved in this process is termed central projection: a point on one plane (the ground surface photographed) is projected onto a second plane (the negative), so that the original point and its image on the second plane lie on a straight line through a fixed point (focal point of the lens) not on either plane. Aerial photographs always involve some distortion, because their vantage point is not exactly vertical and because terrain is seldom completely horisontal or completely flat. Consequently, the information in aerial photographs must be transferred to a ground coordinate system for computation of lengths and areas or before they can be entered in GIS with other map data.
Photographic data are used to create and revise maps, using a variety of photogrammetric instruments. Information may also be verified manually by reference to known details on a map.

6.5   PHOTOGRAMMETRIC MAPPING

Photogrammetry means "picture measurement," and describes a basic process first used in 1851 by A. Laussedat in France. Although Laussedat's work was original, it had little practical application because the lenses and photographic printing techniques of his time were too inaccurate for precision measurements. Close to the end of the 19th century, however, E. Deville, a Canadian, devised a stereoscopic machine to process photographs taken from theodolites. Subsequently, phototheodolites were developed specifically for that purpose. Phototheodolite techniques were eventually widely used, parti-cularly for mapping the Alps of Central Europe. In 1909, the first instrument for map making was built. As in many other fields, it was military necessity that accelerated development. During the first world war, balloons and airplanes were found to be excellent platforms for photography. The airplane eventually proved to be the most successful because a single aircraft could fly along a prescribed course and take numerous overlapping photographs, which were then processed using various stereoscopic machines. The utility of the approach became so dominant that today photogrammetry is regarded as one of the techniques of aerial photography.

The stereo plotter is the instrument most commonly used to transfer aerial photographic information to planimetric and topographic maps which, in turn, may be the bases for vector data.

In a stereo plotter, a photographic interpreter views two mutually over-lapping aerial photographs taken from different positions to form a three-dimensional image. Several systems of forming three-dimensional images are used. Among the older are the anaglyphic systems in which one photograph is projected in red and the other in blue. When viewed by an operator wearing spectacles with a red filter for one eye and a blue for the other, the image acquires the third dimension of height. Another system, illustrated in Fig. 6.19, uses ray rods and an optical system which ensures that each eye sees only one photograph.

In modern analytic plotters, software replaces the rods and all instrument movements are digitally controlled.

By moving a floating point around in the three-dimensional image, the operator can draw in roads, rivers, contours and other details. If the plotter has a digital transmitter, all movements used in drawing in the photo-grammetric model can be numerically encoded and entered in computer storage. Although many analogue instruments, descendants of the original opto-mechanical photogrammetric instruments, are still used, analytical instruments are now the workhorses for GIS data capture.

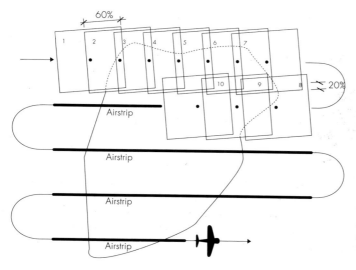

Fig. 6.18
– During aerial photography for mapping purposes, the whole area is covered by aerial photographs usually with a 60% overlap along each flight line and a 20% overlap between flight lines.

Either stream encoding or individual point encoding may be used, as in the equivalent digitiser process. In digital instruments, transformation parameters are used to transform continuously the internal system coordinates to terrain coordinates. The transformation parameters are calculated from the photographic coordinates of points known from field survey control or triangulation. Unlike the digitisation of existing maps, processing photogrammetric data directly encodes elevation information at any one point within the stereo model.

The three-dimensional transformations are:

$$X = c_1 + m\,[a_{11}x + a_{12}y + a_{13}z] \qquad \text{(6-5A)}$$
$$Y = c_2 + m\,[a_{21}x + a_{22}y + a_{23}z] \qquad \text{(6-5B)}$$
$$Z = c_3 + m\,[a_{31}x + a_{32}y + a_{33}z] \qquad \text{(6-5C)}$$

where X, Y and Z are the terrain coordinates, x, y and z are the entered system coordinates and m, c_1 to c_3 and a_{11} to a_{33} are the transformation parameters.

In mapping, new thematic codes and sometimes i.d. codes are entered whenever the category of an object changes, for instance from a residential building to an industrial building.

Photogrammetric encoding requires trained operators and concentration. Experience indicates that i.d. codes are often more efficiently entered on screen in the final editing phase.

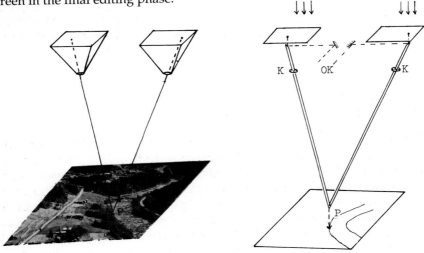

Fig. 6.19 – A stereo plotter; system shown is Wild autograph. a) Aerial photography. b) A three-dimensional reconstruction of the terrain by means of aerial photographs. It is only possible to see the right image with the right eye and the left image with the left eye. When the images are positioned correctly in relation to each other, the brain will merge both image impressions into a three-dimensional model. The light beams in the model can be reconstructed in both analogue and computerised forms. (Ø. Andersen 1980)

The coordinates of such individual points as manholes, drains and property boundaries may be registered in large-scale aerial photographs. So, in the field, such points are often marked with white paint or white paper prior to photographing to ease their identification in the resultant photographs. The size of these markings normally varies from 60 X 60 cm for aerial photo scales of 1:15,000 down to 30 X 30 cm for scales of 1:6,000. Aero triangulation, a special technique for point location, may be used to register marked points with an accuracy of ± 5 μm to ± 10 μm on photographs, which corresponds to 3 cm to 6 cm in an image scale of 1:6,000. For other details, photogrammetric accuracy is 10 μm to 20 μm. (Ø. Andersen)

The product of photogrammetric mapping must be edited to ensure that map data correspond to the map image desired. Map editing programs are used, either during the mapping itself or during the subsequent editing phase. Again, experience indicates that editing can be as time consuming as traditional manual redrawing.

A 60 X 80 cm map in a scale of 1:1,000 can involve one to three MBytes of data, depending on the complexity of the map.

In producing base maps with details and topography, map data normally may be entered at a rate corresponding to the registration of 10 to 12 decares an hour in 1:6,000 scale images, while editing progress may be at a rate of 20 decares an hour. For 1:15,000 images, the corresponding rates are 20 and 40 decares an hour respectively.

GIS cannot distinguish between a digitised map and a digitised stereo model. However, a map is inherently generalised. A photo is not, especially not concerning elevation data.

6.6   POSITIONING SYSTEMS
6.6.1   GLOBAL POSITIONING SYSTEM, GPS

The Global Positioning System, GPS, is a military satellite-based navigational system developed by the Department of Defense in the United States. When fully operational in 1993, GPS will control 21 operational satellites and three reserve satellites. At the time of writing (late 1991), 16 satellites are in operation. Not only are the satellite orbits highly predictable, but the satellites also carry atomic clocks which enable them to transmit highly accurate radio signals. As illustrated in Fig. 6.21, position is determined by using a small, portable receiver to receive and compare the signals from three satellites. For each signal received, the interpretation of codes or measurement of phase permits computation of the satellite's position in space, and the distance between it and the receiver. With these data for each of three satellites, the receiver's horizontally position is computed from the unique intersection of three cones for which the apexes are at the satellites.

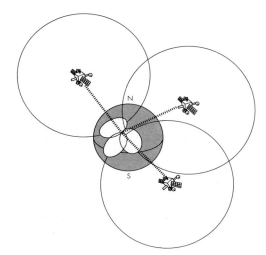

Fig. 6.20 – Global Positioning System (GPS). GPS positions are determined from the intersection of cones.

Horizontally and vertically position may be computed by cross-checking the signals from four or more satellites.

GPS now has two accuracy codes. The most accurate, P-code, is reserved for the military (± 17,8m horizontally, ± 27,7m vertically), whilst the C/A-code is available for civilian uses. Both codes have been available during the test phase of GPS implementation, but the military code will not be available to civilian users after the system is fully operational in 1993. The U.S. Department of Defense is now considering whether to reduce the accuracy of the C/A code, which currently provides positions with an accuracy of ± 28,4 m horizontally, ± 44,5m vertically (± 100 m, ± 156m with reduced accuracy) with no further corrections.

Accuracy can be increased further by simultaneously using data from two receivers, provided that the position of one of them is known accurately. Called differential GPS, this system involves phase measurements and can result in position accuracies of ± 1 to ± 2 cm when the distance between the two receivers is less than 1,000 m. Accuracy decreases when the distance between the receivers increases. In general, differential GPS provides position accuracies of the order of 0.3 m to 2 m for receivers moving at relatively low speeds of up to 20 km/h.

Accuracies such as these are attainable by virtue of extensive, rapid computations of error corrections. In the field, relatively straightforward real-time computations may result in a position being determined to an accuracy of ± 10 m, as for ships in motion.

One of the drawbacks of GPS is that it requires a direct line of sight between the receiver and each satellite accessed. In many instances, particularly for mobile users, the terrain, buildings, trees and other objects may block the satellites from view.

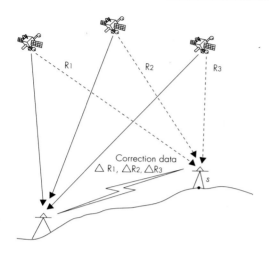

Fig. 6.21 – Differential GPS. Differential GPS are measured in two stations, one of which has known coordinates. Correction values can then be used to calculate the exact position of the unknown point. It is thus possible to calculate the exact position of survey control stations on land or of mobile objects such as boats and cars.

GPS is based on the WGS84 geocentric coordinate system may be transformed to other coordinate systems.

GPS is most useful for:

- locating new survey control stations and upgrading the accuracy of old station positions
- measuring terrain features that are difficult to measure by conventional means
- positioning offshore oil platforms
- updating road data with a GPS receiver in a car
- marine navigation, including integration with electronic charts
- car navigation
- determining camera-carrying aircraft positions to reduce reliance on fixed marks in aerial photography

GPS has been integrated directly with several GIS. In general, GPS data may be regarded in the same way as other digital map data. Coding of GPS data in GIS may be performed in the field to determine accurately the positions of intersections, road signs, the ends of bridges, road surfaces and the like.

The cost of a GPS receiver depends in part on the accuracy it achieves; high-precision receivers, like those used in geodesy, are the most expensive. At the time of writing (late 1991), they are slightly more expensive than electronic theodolites. Low microprocessor costs will certainly boost the market for GPS receivers and spawn new applications, not least systems that combine GPS positioning with GIS route- finding for car navigation.

6.6.2   INERTIAL NAVIGATION

Inertial navigation systems may either support GPS or be used to determine position independently.

The basic inertial navigation system comprises three accelerometers, one on each of the three axes of a platform stabilized by gyro, which measure the triaxial gyro forces. The accelerations measured yield information on triaxial changes of speed which, when integrated, give changes of position in three dimensions. From a start at a known point, an inertial navigation system can compute coordinates as it is moved. If its journey ends at a known point, then its en route computations may be verified or corrected.

An inertial navigation system may be moved at high speed without affecting its accuracy (navigation for military rocketry was an early application). However, all inertial navigation systems drift with time, and must therefore be corrected periodically. The frequency of correction, of course, determines the overall accuracy. Experiments conducted with an internal navigation system mounted on the bed of a small truck driving along a road at 80 km/h indicate that correction every two minutes produces a position accuracy of ± 10 cm. Correction every ten minutes produces a position accuracy of ± 1.0 m (J.M. Becker 1988).

As inertial navigation positionings are independent of light or weather conditions, systems may be mounted in airplanes, helicopters and cars. The latter can be used either to map new roads or as a navigational aid.

Relatively simple inertial navigation systems and compass-like instruments are used in conjunction with GPS when satellite signals are blocked frequently, as with car navigation. Inertial navigation data are used in some GIS, particularly in acquiring road data.

6.7   SATELLITE DATA

6.7.1   OPTICAL REMOTE SENSING

Satellite remote sensing of the Earth began in the 1960s, when the technical capabilities of satellites operating in Earth orbit converged with the increasing ability of computers to manipulate large quantities of data. Two observation modalities evolved: passive optical, which deals with reflected sunlight and re-emitted thermal radiation, and microwave which, like radar, deals with the transmission and reflection of energy in the microwave portion of the radio-frequency spectrum. Apart from the nature of the phenomena observed, the difference between the two modalities is technical. The optical spectrum is considered to extend from the very short wavelength ultra-violet region to a wavelength of 1000 μm in the longer infra-red region. This is the short-wavelength limit to radiation which can be conveniently generated and detected by microwave electronics devices. In mid 1991, ESA – the European

Space Agency – launched the ERS-1 Earth Resources Satellite with a payload containing radar imaging instruments. At this writing, the ERS-1 has not been operational long enough for its data to have impact on mapping the resources of the Earth.

At the time of writing (late 1991), the two major earth resource scanning satellite systems in operation are of the passive optical type. LANDSAT is the generic name for a series of five satellites launched between 1972 and 1984 by the United States. SPOT, an acronym for Systeme Probatoire d'Observation de la Terre, uses a satellite launched by France in January 1986. Both systems are used primarily for mapping vegetation, geological features and soil types, and to some extent for water parameters and linear structures such as roads and rivers. SPOT can register three-dimensional images for processing to yield gross elevation data.

At any one instant, a satellite carrying sensors that perceive reflected solar energy views a scene on Earth. In principle, the sensors might perceive energy at wavelengths anywhere in the optical region of the electromagnetic spectrum, from 0.4 µm to 1000 µm. Traditionally, the optical region is divided into sub-regions, each of which is termed a spectrum:

Sensors are available for all spectra, from ultraviolet to far infra-red, but not all may be used in optical remote sensing. This is because not all portions of the total optical spectrum are transmitted equally through the atmosphere; many are absorbed, some almost completely. So the variations of atmospheric transmission, and absorption with wavelength, limit the optical spectrum that is useful in remote sensing to transmission bands, or windows, in the range of 0.4 µm to 15 µm.

Sensors on satellites operate in a similar way to those in camera light meters, but instead of measuring energy only in the visual spectrum as a light meter does, together they may be arranged to measure energies in several spectra, from the visible through to the mid infra-red. Multispectral scanning,

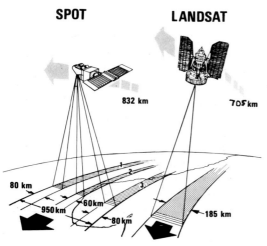

SPOT   **LANDSAT**

832 km   705 km

80 km   950 km   60km   80 km   185 km

Fig. 6.22 – Satellites in orbit. "Mapping" satellites trace a continuous track over the earth's surface whilst circling the earth. The reflected electromagnetic radiation from each swath is registered as digital values from which images can be constructed. a) SPOT b) LANDSAT (NOU 1983:24 Satellitfjernmåling)

as it is termed, results in Earth image data in each of the windows scanned. The spectral windows are chosen to monitor the wavelengths of greatest interest.

Waters, woods, cultivated fields and other areas are distinguished by characteristic reflected and re-emitted energy spectra, which are recognised in the satellite images.

Spectrum name	Wavelength range
ultraviolet	shorter than 0.4 μm
visible	0.4 μm to 0.7 μm
near infra-red	0.7 μm to 1.0 μm
solar reflected infra-red	1.0 μm to 3.0 μm
mid infra-red	3.0 μm to 15.0 μm
far infra-red	longer than 15.0 μm

Fig. 6.23 – Spectrum classes.

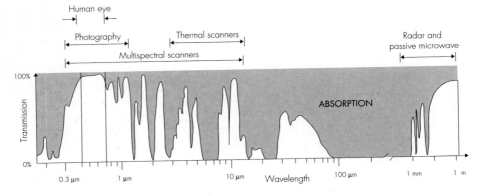

Fig. 6.24 – Electromagnetic spectra

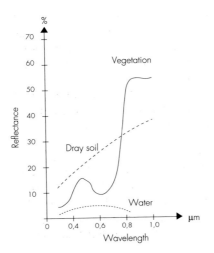

Fig. 6.25 – Typical spectral reflectance curves for vegetation, soil and water.

Sensor LANDSAT (4,5)	Band	Wavelengths, μm			Resolution, m.	Ground-swath width 185x185 km
MSS	4	0.5 –	0.6	(green)	80	
	5	0.6 –	0.7	(red)	80	
	6	0.7 –	0.8	(near IR)	80	
	7	0.8 –	1.1	(near IR)	80	
LANDSAT (4,5)						185x185 km
TM	1	0.45 –	0.52	(blue)	30	
	2	0.52 –	0.60	(green)	30	
	3	0.63 –	0.69	(red)	30	
	4	0.76 –	0.90	(near IR)	30	
	5	1.55 –	1.75	(mid IR)	30	
	6	10.4 –	12.5	(thermal IR)	120	
	7	2.08 –	2.35	(mid IR)	30	
SPOT	Multispectral	0.50 –	0.59	(green)	20	60x60 km
		0.61 –	0.68	(red)	20	
		0.79 –	0.89	(reflected IR)	20	
	Panchromatic	0.51 –	0.73	(green)	10	60x60 km

Fig. 6.26 – Characteristics of LANDSAT 4.5 and SPOT

The images are compiled successively along a track, or path, of satellite motion as projected on the surface of the Earth. Each sensor in a satellite comprises numerous detectors, each of which registers energy reflected from a small square on the surface of the Earth. Data from the squares are stored as image elements (pixels) in the satellite, with digital values corresponding to radiation intensity. Together, the sensors perceive a cross-track swath, with the motion of the satellite along the track generating the raster images of the surface scanned. Data are transmitted via microwave link to Earth stations.

The SPOT satellite orbit is polar (passing nearly over the poles) at an altitude of 832 km. The satellite is solar synchronous; that is, once every 26 days it passes over the same point on Earth at the same time of day. In other words, it scans the entire Earth every 26 days. SPOT has a two-sensor system. One sensor registers the visual spectrum with a resolution, called its instantaneous field of view (IFOV), of 10 X 10 m. The other is a multispectral sensor that registers three channels (green, red and infrared) with a resolution of 20 X 20 m. A single satellite image, or scene, covers an area of 60 X 60 km.

The orbits of the five LANDSAT satellites are also polar and solar synchronous, at altitudes of 918 km for the first three satellites and 705 km for the other two. Each satellite scans the entire Earth every 18 days. The LANDSAT 4 and 5 satellites have two types of sensor. One, the Thematic Mapper (TM) sensor, registers seven channels in the visible and infra-red spectra with an instantaneous field of view (IFOV) of 30 X 30 m, except for the thermal infra-red channel which has an IFOV of 120 X 120 m. The other, the Multi Spectral Scanner (MSS) sensor, registers four channels with an IFOV of 80 X 80m. A single LANDSAT scene covers 185 X 185 km.

Because they perceive reflected solar energy, almost all the SPOT and LANDSAT sensors fail to "see" through cloud cover or in darkness. The single exception is the LANDSAT mid infra-red spectral band sensor which perceives re-emitted energy. Therefore it is often referred to as a thermal infra-red sensor.

Radiometric Correction

The measurement magnitude, or radiometric value, of each pixel may be degraded by noise caused by:
- sensor inaccuracy
- atmospheric attenuation and scattering of reflected energy
- variation of reflected energy as solar incidence varies with the curvature of the Earth
- variation of reflected energy due to changes of surface slope

As most of these causes of pixel noise cannot be quantified directly, radiometric corrections are based on assumptions and models. The goal is to apply corrections so that the pixel values representing any one phenomenon will be the same, regardless of their locations in the overall image.

Geometric Correction

Geometric correction is necessary if the data generated are to be used in mapping or for performing various analyses. The geometrical accuracy of data are influenced by:
- rotation of the Earth during sensing
- curvature of the Earth
- variations in surface elevation
- satellite instability in orbit
- instability of recording devices
- inaccuracies in the sensor projection system
- atmospheric aberrations

Geometric correction of satellite data may be performed in various ways at two levels of precision:

Level 1
Systematic errors due to
- rotation of the Earth
- instabilities of the satellite and its instruments
  (panorama effect, striping, angle of observation)

Level 2
- internal image improvement
- transfer to chosen cartographic projection
- reduction to mean terrain elevation
- transformation based on map coordinates
- corrections for terrain variations

After all corrections have been made, geometrical accuracy may actually be better than resolution. For instance, data with a resolution of 10 X 10 m may be geometrically accurate to ± 6 m to ± 7 m (Satellitbild), and data with a resolution of 80 X 80 m may be geometrically accurate to ± 50 m to ± 70 m (T. Bernhardsen, 1975). This implies that these data are best used for overviews of areas that were previously poorly mapped, such as in developing countries.

6.7.2   ANALYTIC METHODS

As a rule, satellite images must be analysed and interpreted to yield the information desired. Satellite data are digital, so analyses and interpretations may be partially computerised.

Three analytic methods are now common:
- colour enhancement
  - RGB colour adapting
  - HLS colour adapting
- classification
  - supervised classification
  - unsupervised classification
- segmenting

In practice, these three methods are often used together.

Colour Enhancement

Satellite image data have no colours in the sense that the eye perceives colour in the visible spectrum. All colouring of satellite data relayed to Earth is done after the data are compiled into images. Normally, the data are processed interactively on screen in order to compile colour images that enhance the themes of interest in mapping, such as coniferous and deciduous forest, cultivated fields and so on. As the colours do not need to be the same as those seen by the unaided eye in daylight, the images created need not resemble conventional colour photographs. In fact, in most cases they do not.

Colours are assigned to the individual spectral bands perceived by a satellite sensor assemblage in order to yield a false colour enhanced image. For instance, red, green and blue colours may be assigned to three LANDSAT spectral bands to produce an image in which deciduous trees appear orange, coniferous trees brownish, earth and rock blue, and water black.

The basic colours may be adjusted individually, although the disadvantage of the simplest approach – basic colour adjustment – is that adjusting any one colour affects the aggregate image. For instance, amplifying the red basic colour in an image not only makes the red pixels redder, but also adds yellow to the green pixels.

Often, a more suitable method of improving false colour enhanced images is to adjust the colour hue, lightness and saturation (HLS). In any event, the final colour enhanced images may be interpreted in the same way as aerial photographs.

Classification

Classification is based on the assumption that like phenomena in the image belong to the same class on the ground, and have similar characteristic spectra; that is, they have a "spectral fingerprint."

There are two approaches to classification that are distinguished primarily by their initial assumptions. In supervised classification, ground truth data from direct in-field observations are used to identify the initial parameters used in classification. Unsupervised classification needs no ground truth in its initial stages, but the final images and maps produced must almost always be verified in the field.

Supervised Classification

Supervised classification begins with a compilation of training data. Firstly, each class to be mapped, such as water, cultivated fields, pine trees, fir trees, and so on, is named and assigned a colour. Then, a skilled analyst works on screen, circling areas known to contain the classes of objects named. Spectral band means and deviations are calculated for each area encircled. In other words, ground truth data are used to "supervise" subsequent manipulations.

Once the training data are compiled, the analyst initiates automatic classification of all classes with ground truth in the entire image. Each pixel is classified according to the greatest probability of its belonging to a particular class. Hence, the entire image may be classified using relatively few in-field observations. Various statistical methods are used to assign pixels to classes, including contextual classification which takes account of the values of neighbouring pixels.

The pixels of the classified image displayed on screen are coloured according to the class colours assigned in the training data.

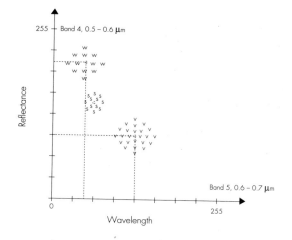

Reflectance

255 ┼ Band 4, 0.5 – 0.6 μm

Band 5, 0.6 – 0.7 μm

255

Wavelength

Fig. 6.27 – Clustering. Clustering is the process whereby pixels which are adjacent in the spectrum are grouped together. Several parts of the spectral bands can be used simultaneously.

w = water; s = soil; v = vegetation.

Unsupervised Classification

Supervised classification requier ground truth data when starting the classification. The classifiaction is also limited to the classes with ground truth. Unsupervised classification is used to overcome these obstacles.

Unsupervised classification starts with an unsupervised cluster training procedure. The first step is clustering, a way of ordering data by sorting pixels into classes according to their spectral values. In other words, clustering comprises classification by spectral characteristics rather than by relation to ground truth observations. The analytical procedure employed may be interactive and includes four main steps:

1. Each cluster is assigned a value corresponding to its centre.
2. Every pixel in an image is placed in the nearest cluster most closely matching its value, according to a method of distance measuring.
3. When all pixels of an image have been assigned to clusters, new centres are calculated to form the means of the constituent pixels.
4. Repeat steps 2 and 3.

The process stops when the clusters no longer change between successive iterations. The clusters are then labelled by assigning colours and in-field observations used to liken the coloured clusters to ground phenomena.

Segmenting

Segmenting is usually performed automatically by a computer program. It starts with the comparison of a pixel value with the values of its eight immediately neighbouring pixels. Of these eight, those with values – within set limits – that are closest to that of the start pixel are selected. These pixels are then compared to their neighbours, which are also selected if their values are within the set limits. The selected pixels form a segment. The process is repeated until no more neighbouring pixels fulfil the criterion for selection. At this stage, a new start pixel in a new area is chosen, and a new segment is successively built up. Segmenting thus takes account of both geometry and pixel values in grouping pixels.

Integrating Satellite Data in GIS

The data of a geometrically-corrected satellite image that were acquired via a satellite spectral band may be regarded as a thematic layer of pixels. Multispectral data can therefore constitute a GIS overlay. Images taken at other times or by other satellites may be adapted to the same map projection, coordinate system and pixel size, and thereby form a new GIS overlay. Map overlays in raster form (geology, vegetation, etc.) may be integrated in the same system provided their respective geometries and map references are the same as those currently used in the system.

The classified raster data may be converted to vector data by using suitable conversion functions. The data may then be entered in vector GIS.

6.8   TEXT DATA

Public administration agencies, public utilities and other organisations dealing with georeferenced data may frequently need to integrate text data in GIS. All kinds of documentation may be involved – from letters and deeds to zoning directives and service overviews.

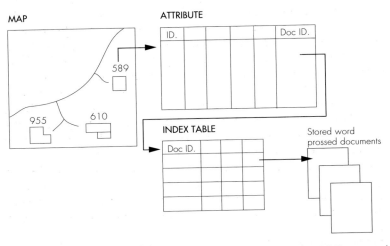

Fig. 6.28 – Documents in a word-processing system can be integrated into GIS via an index table between the attribute table and word-processing system.

Text data may be integrated into GIS via a table of object identifiers and pointers to text file addresses. A GIS may also conduct searches of text documents. In practice, this requires careful organisation in the naming and filing of text data.

GIS graphic map images may be imported in word processing and incorporated in texts using programs such as MS Windows or PostScript. Texts with maps may then be printed out on a laser printer.

6.9   COMPUTER AIDED DESIGN (CAD)

CAD-generated drawings may be integrated into GIS in the same way as texts, via tables with pointers to CAD file addresses. Or they may be entered directly, as with map data, via a common transfer format.

6.10   ATTRIBUTE DATA

Just as digitising work may be divided into three phases, so may the acquisition of attribute data be divided into:

- preparation
- entering
- editing

Preparation

Data of interest are often recorded on paper and scattered amongst a multitude of different conventional registers and files. Consequently, the initial acquisition of data may require a major organisational effort jointly involving many different agencies. For instance, registering all undeveloped land in a county may require data from all the agencies in all the townships in the area.

In some cases, registration has been organised in accordance with statutes and regulations. For instance, in Norway townships are now required to report all new land ownerships, addresses and buildings to a central database, named GAB, an abbreviation for Grunneiendom, Adresse, Bygninger, the Norwegian terms for the three categories of data filed. In other cases, it may be necessary to supplement new field registration with local knowledge.

Attribute data may be collected effectively only if a suitable collection structure is available, or is specific as an initial step. A common approach is to complete various forms, either on paper for manual registration and subsequently entry via a PC, or to hold them in files for direct keying in on a laptop/PC.

The data fields of the forms should, of course, correspond to the data fields of the database. Registration should not be ambiguous and should provide for all data variations to ensure that data always locates in the correct data fields.

Consider, for example, setting up a GIS for a water supply system. The objects specified may be as follows:

Piping	1. Water mains
	2. Supply mains
	3. Service pipes
	4. Sewer pipes
	5. Venting
	6. Manholes
	7. Drains
	8. Inoperative piping
	9. Inoperative piping not to be removed
	10. Basins
	11. Booster stations
	12. Pump stations
	13. Pressure zones
	14. Stop valves
	15. Joints and joining methods
	16. Non-return valves
	17. Pressure reduction valves
	18. Separators
	19. Vent valves
	20. Fire hydrants, above ground
	21. Fire hydrants, below ground

Service pipes	1. Type
	2. Tapping
	3. Sprinklers
	4. Stop valves
	5. Water meters

Manholes	1. Type
	2. Piping accessed
	3. Status (operational, planned, abandoned)
	4. Location
	5. Construction

Miscellaneous	1. Units of length	6. Soil conditions
	2. Dimensions	7. Road traffic conditions
	3. Model numbers	8. Damage
	4. Valve numbers	9. Repairs
	5. Supply areas	10. Service interruptions

Fig. 6.29 – A manhole. There are many details in this picture of a manhole which have to be registered for entering in a GIS.

These data may be divided roughly into three categories: locations, components and network parameters.

The acquisition of attribute data often requires verification in the field. Verification may entail going down a manhole to sketch or photograph details; rodding pipes to ascertain direction and junctions; measuring and levelling; poring over old files and drawings to locate details of dimensions and materials; interviewing retired plumbers, and so on.

Once these data have been collected and structured in lists, on forms, and/or on maps, entering data in GIS may begin.

Entering Data

One of the more expedient ways of entering data in GIS is use a database application program on a laptop/PC for initial entry, and subsequently export the database file to GIS. PC database application programs support forms that can be designed and filled out on screen, so a specific form may be designed for each type of object.

Form design should include verification of formats in all data fields to ensure, for instance, that text is entered into a text data field whilst only figures are entered in a numerical data field, and that entered values are restricted within set extremes.

Occasionally, it is advantageous to call up a graphic display of map data on screen as the attribute data are entered.

It may also be useful on-site to microfilm information contained in old files. Processed microfilm may be viewed on a microfilm reader placed beside a PC/terminal keyboard, so data may be keyed in as the microfilmed pages are viewed. The land ownership part of the Norwegian GAB database was built up by using microfilm records of the deed registers of all the Land Registry offices in the country. This circumvented the need for manual form-filling. Nonetheless, the aggregate time involved in entering data on 2.2 million properties amounted to 150 man-years, spread over a period of 5 years.

Whenever data are available in computerised form, they may be entered directly.

Editing

Automatic verification functions reveal only formal errors. Incorrect information, such as a wrong name entered in a name field, can be spotted only by manual copyreading. Major and meaningless errors are usually easy to spot, but incorrect spellings, inversions, omissions and other less obvious errors are more difficult to find.

Consequently, it is better to verify attribute data by printing out sorted lists which may then be checked line by line by a copy reader.

The elimination of errors in entered data is important, although it is invariably time consuming because it mostly requires manual work.

Once the digital map data and attribute data are verified, corrected and entered in GIS, the identifiers linking the databases should be checked and all missing identifiers marked. Checking in GIS will not identify incorrect links due to map data code errors, or mistakes (in legal fields) in attribute data entered.

6.11   IMPORTING COMPUTERISED INFORMATION TO GIS

6.11.1   BACKGROUND

Most countries administer databases containing geographic information; in some countries, such as Denmark, as many as 400. This means that it is often necessary to import database map and register information into GIS, and to transfer data between different GIS facilities.

However, when the storage format differs from the application format, export-import data transfers can be complicated, which is why many GIS can import and export data in a variety of formats.

If a read routine is unavailable for a particular storage format, it may be necessary to compile a transfer program. Indirect methods may also be used, for example a transparent transfer, in which the stored data is first imported to a compatible GIS with the requisite reading ability, and then reformatted and transferred to the target GIS. In effect, the intermediate GIS is transparent to the data flow from the database to the target GIS, although some interchange of protocols may be required.

Imported map data should include all graphic information to ensure that plotted maps are identical to the originals. Graphic information includes individual points with their thematic codes; symbol numbers and rotation of symbols; open and closed polygons with their thematic codes; line types; elevations of points in polygons; arcs with their thematic codes; radii; directions of flow and texts with their fonts.

Often, pure map data may be stored as spaghetti data, which must be edited to topological data if it is to be useful in GIS.

In addition to the details of the data formats, one needs to know the particulars of the transfer medium (such as magnetic tape), the data administrative details (data volume, number of lines, etc.), block sizes, record length, data structure and the meaning of each data field.

Attribute data or tabular data may be transferred relatively easily between relational databases. The transfer of data from hierarchical databases or network databases to relational databases may be considerably more difficult, however, because the respective structures differ.

6.11.2   STANDARD TRANSFER FORMATS

In many countries, standard transfer formats for vector and raster geographic data have either been developed, or are currently being developed.

Transfer formats usually involve file formats dedicated to transferring large quantities of data – a few hundred thousand characters or more.

The major elements of file transfer formats include syntax, a conceptual model and a code set. This is regardless of whether larger units, such as entire map sheets, are transferred, or databases are being established or updated. Updating entails the full deletion of data from an area and full re-entry of the updated data, which is extremely inefficient. So a need has arisen for object-oriented formats which include mechanisms for dynamic updating of individual objects. Thus far, few geographic formats have this capability.

The next level of standards involves distribution via communications lines and networks. This is known as Electronic Data Interchange, or EDI. The standards evolved should therefore include stipulations of message level.

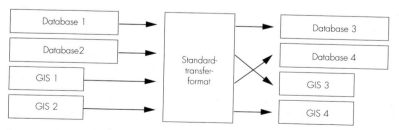

Fig. 6.30 – By means of standardised exchange formats, digital data can be exchanged between different GIS and also between GIS and different data bases.

Effective transfer also requires standardised data structures and common data models. Both sender and receiver must not only use the same language, but also must have a common basic understanding of the content and meaning of the data involved.

At the time of writing (late 1991), the European Transfer Format, ETF, is being developed in Europe. Based on the European Territorial Database, this is a database for small-scale maps. Meanwhile, the National Transfer Format, NTF, which handles both vector and raster data, is being developed in England, whereas Finland has accepted the EDIFACT standard, which combines several needs. In 1989, a recommendation for a thorough, theoretical standard, Spatial Data Transfer Standard, SDTS, was published in the U.S.A. A topologically structured format, Digital Line Graph, DLG, developed by the U.S. Geological Survey, is in use in the U.S.A. Germany has an ambitious program in ATKIS (Amtliches Topographisch-Kartographisches Informations-system), which includes both a topological data model and a cartographic data model. NATO's DIGEST format for vector and raster data has attracted interest in the civilian sector.

The requirements for hydrographic data transfer formats are strict. The DX90 format, which supports both format transfer and dynamic updating of individual objects, is being developed under the auspices of the International Hydrographic Organisation (IHO).

In addition to these standards, pro forma industrial standards for plotting general graphic data have also been developed. These, too, can be used for transferring digital map data. Today, the most popular standards are AudoCad's DXF format, Intergraph's SIF format, and Hewlett-Packard's HPGL format. PostScript, which is a graphic language, also has potential in plotting, printout and transferring of geographic data.

Data volume often depends on how data are formatted and designed. General formats must provide leeway for all possible situations. Consequently, in practice the data often include basic administrative data which are not used, or space for attribute data which is not used.

6.12 DATA STRUCTURING

Data collected and entered in a database must be processed and arranged to provide information that is meaningful for current tasks as well as for future tasks still to be defined. This information might include the answers to questions relating to

- location of a particular object
- attributes of a particular object

- attributes of all objects in a region
- identification of regions containing objects with attributes A and B
- quickest route between points A and B
- identification of all objects of type A contiguous with objects of type B

In order to be useful, data must be organised. Disorganised data are unproductive: in other words, rubbish in, rubbish out. For instance, if the intention is to show all the water pipes with a three-inch diameter in a network, the result will be of doubtful value if the pipe data are sometimes entered as inside diameters and at other times as outside diameters. A transport analysis can be performed only if elements of the road network – its bridges, tunnels and ferry routes – are entered as secondary parts of the primary network, and not as independent elements. If all the properties on a specified part of a street are to be identified, the data must contain stable, common identifiers between streets and properties. Similarly, if all the private roads of a township are to be identified, the concept of what constitutes a private road must be clearly stated.

Clearly, then, data planning and structuring are essential to successful use in GIS, and almost all GIS projects should begin with careful analyses of how data are to be entered.

The data model and its related structuring must be independent of the chosen software and hardware. In many ways, the problem is the converse of selecting software and hardware to suit the data model.

Data may be structured using the following procedure:

1. Defining areas of application
   Firstly, the real world situation must be limited to the area of application, e.g. the operation and maintenance of water supply and sewage networks, property management, transportation planning, and so on. Secondly, if the database is to contain several areas of application, each must initially be treated separately.

2. Constraints
   The tasks involved and the products to be produced are then identified for each area of application. Examples of constraints include:

   Tasks
   • administrative and operational tasks
   • statutory public tasks
   • planning
   • design tasks
   • dissemination of information

Requirements
- requirements and needs for standardised map products
- requirements and needs for standardised reports

3. Criteria for assigning priorities
   As each constraint identifies a number of object types with their attendant attributes, only relevant data should be entered in the database and the numbers of object types must be constrained. Thus, the data on a road surface are important, but not as important as the width of the road, which, in turn, is not as important as its geometry and location. Objective criteria must then be chosen to determine the objects included. Here, cost-benefit analyses are often a good starting point.
   In practice, however, the problem is often the converse: how can the necessary data be acquired?

4. Descriptions of object types and attributes
   The chosen object types and their attributes must be described. Each type of object comprises a collection of individual entities to be treated equally, given names and defined. For instance:
   - object type: road
   - definition: all public and private roads passable for cars, excluding detached roads and byroads shorter than 50 m.
     Attributes are unique descriptions of objects. Each attribute has a definition and a permitted range of values. For instance, the attribute description for the object "road" may be:
     – attribute: load capacity
     – permitted range: minimum 0 kg, maximum 20 tons

5. Coordinating terms and definitions
   The various object terms and definitions must be coordinated, particularly for common data. Disparities, of course, are inevitable. For instance, whilst parts of roads may be regarded as property boundaries in property administration applications, they would certainly be defined as roads in road and transport applications.
   Attribute overviews must be uniform. All attributes for a single object type, such as for "roads," should be assembled and assigned "globally permitted" values.

6. Geometric representation
   Rules must be established to determine how objects are to be represented geometrically and which basic geometric elements are to be used. In principle, the choice is between:
   - vector representation (points, lines, areas)
   - raster representation

Entities comprising several objects must be allowed.
In principle, the areas of application determine how objects are to be represented, but other elements, such as costs and ease of updating, must also be considered.

7. Relations
   Relations between objects must be defined before tasks can be addressed. Relations may include:
   - composition of objects
     e.g. a country comprises counties, a county comprises townships, and a township comprises statistical units
   - locations of objects
     e.g. a particular building is located on a particular property, or a planned development is in a recreation area
   - affiliations of objects
     e.g. which addresses are particular to a street, which stop valves are parts of which water pipes
   - neighbours of objects
     e.g. which properties border on a particular property, which type of land use borders on a preservation area

   Some relations may be computed from digital map data, as when forming topology, whilst others must be entered as attributes. Sometimes it may also be necessary to distinguish between actual relations and potential relations. Concrete user needs should dictate the selection of the relations described. Relations that are not used need not be described.

8. Quality requirements
   The data must be subjected to quality requirements, for instance:
   - geometrical accuracy
   - attribute accuracy
   - geometrical resolution
   - consistency of linking between geometric data and attribute data
   - currentness
   - comprehensiveness with respect to content and geographical coverage

9. Coding
   Code lists of the geometric object type designations must be compiled. The code system should be based on the data structure defined in steps 4, 5 and 9 above.
   Code lists may also be compiled for attributes and for descriptions of relations.
   The identifiers between geometry and attributes must be designated

and, if possible, their coding should be delineated.

The completion of steps one to 10 results in an object catalogue which contains descriptions of all the objects to be entered in the database. Once the data have been structured in a theoretical model, it remains to:

- realise the logical structure in a GIS with a concrete database application system
- choose the physical files to be structured on disks for optimum utility

Experience indicates that structuring entails problems, of which the most common are that:

- existing maps are not always good data models
- terminology is imprecise and not based on common concepts

Maps are basically good sources for describing objects and their attributes. However, maps always represent particular models of the real world, and GIS should represent the real world, not the maps that depict it. For instance, ferry routes are often shown by dotted lines on maps, whereas in data models on transport planning they should be integral parts of a contiguous road network.

Uniform terminology may also be difficult to achieve. For instance, does a pedestrian area that is accessible to emergency vehicles classify as a road?

6.13 FORMAL PROBLEMS IN ESTABLISHING GIS

Use of geographical data often poses the question of who owns the data. Laws and statutes usually regulate who may claim property ownership. Those who can claim property ownership may also regulate property use through taxation and other price mechanisms.

The legal status of geographic information is not always clear. In some countries, geographic information is regarded as a work of art, like a painting or a novel, and the ownership of it is regulated under copyright law.

In most cases, computerised maps and registers have been generated in the public sector. The question of access to the data then depends on whether public officials regard geographic data as belonging in the public domain or as a product protected under copyright law. Practice varies considerably from country to country, and sometimes within a country, as from state to state in the U.S.A.

In England, the public sector claims copyright. Hence, permission must be granted by the Ordnance Survey (the national map agency) for all digitising of existing maps. On termination of the digitising, a copy of the resultant digital data must be provided free of charge to the Ordnance Survey. In addition, the Ordnance Survey levies a charge proportional to the amount of digitising work carried out.

In Canada, the public sector also claims copyright over map data, but not as rigorously as in England. Usage fees are charged whenever the value of the digitised data exceed the costs incurred in acquisition and digitising.

The business aspects of GIS have yet to be developed fully, primarily because so far most GIS facilities have been established to meet the needs of particular public agencies. However, commercial uses will undoubtedly become more important in future. The development of electronic marine charts and systems for car navigation will require equipment suppliers to deliver digital map data together with the equipment sold. Databases may then become commercial wares with considerable profit potential. In that event, data ownership will become an issue.

National geographic databases and GIS may be regarded as parts of future infrastructures, as with road networks and electricity supply networks. In the United States, governmental maps and digital map data are regarded as belonging in the public domain, and all copying and further uses are free of charge. That view is apparently based on the belief that society on the whole profits from more valuable products and data that initially were financed by the taxpayers.

In many countries, use is regulated by price mechanisms. Thus, the Norwegian Mapping Authority sets lower usage fees for the public sector's use of data and higher fees for commercial uses in the private sector. Correct pricing, however, is difficult. High prices encourage users to compile their own digital map data, which then reduces the importance of the public sector. Low prices can result in data not being kept up to date.

Many countries have laws pertaining to the establishment and use of registers of persons. Official permission is almost always required for such registers, a restriction imposed as a means of safeguarding individual rights. Geographic data which contains data about individuals may similarly be restricted.

Sensitive data, such as maps of military areas or data concerning endangered species or rare plants may be considered of national interest, and restricted accordingly.

In some cases, combining data from open registers may be illegal. Limited information may be openly available, but when many data sets are combined to cover larger areas or effect undesirable combinations, the result may constitute a security risk. For instance, for national security reasons, tele-communications administrations prohibit unauthorised persons from hand-ling the network information contained on several map sheets. In general, the more detailed the information, the stricter the constraints on its use.

Access to computerised information may also be restricted because those who compiled it may not be confident of its quality and therefore prefer that it should not be widely used. Examples of this include data gathered ad hoc for special projects and not based on sound methods or quality controlled.

7 DATA QUALITY

7.1 SELECTION CRITERIA

Users frequently expect digital data to be of higher quality than conventional map data or data in manual registers. This is both because digital data have a far broader range of application than analogue data and because users presume that technological advances also enhance the quality of data.

The applications of digital data are indeed many and varied, but digital data systems are not inherently more accurate than the analogue systems they supplant. In general, digital systems are capable of processing data more precisely than analogue systems, but their overall accuracy still depends on the accuracies of their source data, which in most cases remain analogue.

To see why this is so, consider time, perhaps the most accurately measured quantity in everyday life. In the 1950s, extremely accurate ship's chronometers and other timepieces became available, at high prices; common wristwatches and clocks were accurate to only a few minutes a day. In most applications, clock accuracy was a question of how much a user was willing to spend for a finely-made mechanical timepiece. National standards of time were kept by using large electronic clocks which derived their signals from the oscillations of quartz crystals housed in temperature-controlled ovens. These clocks were completely analogue and accurate to the order of one part in 109, or about one second in ten years. Thanks to microchips and digital techniques, similar levels of accuracy are now common in ordinary wristwatches and clocks. The accuracy is still due to quartz crystals, although it is the digital techniques (and their ultra-compact realizations using microchips) that have made such accuracy available in small timepieces. In terms of overall accuracy, the quality of a timepiece now depends both on the quality of the crystal oscillator, and on the precision with which its signals are processed to give the final display of time. In purely technical terms [marketing oversell of expensive timepieces notwithstanding], user needs determine accuracy. In general, accuracy follows price because greater accuracy requires a higher quality crystal (or atomic) source and more complex processing of its signals.

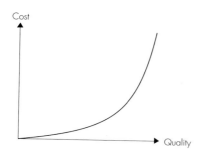

Fig. 7.1 – Relationships between quality and cost.
There is a strong correlation between data quality and cost – costs increase with quality. Quality is often decided on the basis of cost.

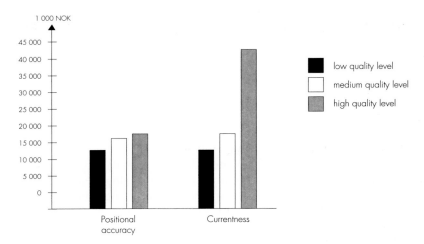

Fig. 7.2 – This figure shows a simulation of cost sensitivity in relation to different quality standards for the national road data base for Norway. Is is apparent that currentness is here much more susceptible to quality choice than positional accuracy. See figs. 11.19 and 11.20.

In GIS, the situation is similar: final accuracy depends on source accuracy. In principle, the user dictates the need for greater accuracy. This is then provided to some degree by improved technology, which is purchased or devel oped in response to the original need. But although technical achievements are virtually limitless from a technical viewpoint, the actual level of achievement, and thus the available technology, is determined – and consequently limited – by costs. The final accuracies of data depend on the qualities of the original input data and on the precision with which input data are processed through all the various stages. Higher accuracy entails higher initial data quality and more precise processing, both of which increase system costs. Furthermore, the requirements for currentness affect the measure of data quality and hence system cost. In general, the costs of updating may strongly influence the determination of data quality.

So, in theory, GIS data quality is a compromise between needs and costs. However, in practice the choice is often a question of shopping for what is currently available or can be acquired within a reasonable amount of time.

Delays, of course, may be due to a variety of causes. In countries where snow falls in winter, users of aerial photographs must wait until the snows melt before they can proceed. Similarly, users of satellite images may have to wait for cloud cover to disperse over an area of interest. Or users of map digitisations produced by national mapping agencies may be assigned to backorder queues because the speed of agency work is often constrained by budget allocations. Finally, users may suffer delays in resolving copyright or other legal difficulties concerning the ownership of data. Any of these factors may cause data to be of lower quality than originally envisioned.

In summary, four aspects of data acquisition comprise the criteria for selecting data accuracy:

- needs
- costs
- accessibility
- time frame

The most important measures of data quality are:

1. georeferencing accuracy
2. attribute data accuracy
3. consistency of links between geometries and attributes
4. geometry link consistency (topology)
5. data completeness and data resolution
6. data currentness

7.2   MEASUREMENT ACCURACY AND PRECISION

Accuracy is customarily stated in terms of the standard deviation of measurements, or root-mean-square deviation as it is known to mathematicians and engineers. The standard deviation for discrete samples extended to control measurements of selected observations is given in Eq. 7-1.

$$m = \pm \sqrt{\frac{\sum_{i=1}^{n} V_i \cdot V_i}{n}} \qquad (7-1)$$

where $V_i = X_i - X_i$, X_i is the control measurement value of the i'th observation, Xi is the observation in question, and n is the number of control observations.

Deviations may be random or systematic. The systematic errors may be computed according to Eq. 7-2.

$$s = \sum_{i=1}^{n} (V_i) / n \qquad (7-2)$$

The systematic component may be relatively unimportant when data originating from the same source are used, but it can be disruptive in data originating from dissimilar sources.

Data accuracy may be verified through reference to other, more accurate data that have been designated as error-free for the purpose. These reference data may be compiled from survey measurements in the field or from maps of considerably larger scale.

Errors may accumulate in data processing. For independent error sources, the median error of a summation C = A + B is as given by Eq. 7.3.

$$m_C = \pm \sqrt{m_A{}^2 + m_B{}^2}$$

$$(7\text{-}3)$$

Precision and Accuracy

Although they are commonly considered to be identical, precision and accuracy express different aspects of measurements and therefore should be carefully distinguished from each other. Accuracy expresses how closely a measurement represents the quantity measured. Precision expresses the repeatability of the measurement. For the measurements normally of interest in GIS, precision is limited only by the instruments and methods used. As these improve, accuracy improves. In physical measurements, precision is customarily indicated by the number of significant figures carried in a quantity expressed. For instance, a figure of 101.23 cm indicates that distances of the order of one metre can be measured with a precision of one-tenth of a millimetre. Accuracy is usually stated in terms of an interval in which a true value is assumed to lie. For instance, 101.21 ± 0.04 cm indicates that the true value is assumed to lie between 101.17 cm and 101.25 cm.

7.3   POSITIONAL ACCURACY

The goal of accurate georeferencing is to locate objects exactly as they are located on the ground, as related to a common coordinate system.

In ordinary mapping, accuracy is inversely proportional to map scale: a map in scale 1:1000 is more accurate than one in scale 1:100,000. The corresponding case for digital map data is more complex. Dissimilar data are stored in the form of terrain coordinates, regardless of their respective sources and accuracies. In other words, all data are stored in a scale of 1:1 and may be used to generate maps of any scale.

The use of insufficiently accurate data may eventually cost more than the original acquisition of the data. A GIS can plot smooth lines on a map and perform convincing computations and analyses, all on the bases of its stored data and regardless of the accuracy of the original data entered. Moreover, a GIS can perform calculations of cut and fill works from data accurate to ± 2 m as comprehensively as it can from data accurate to ± 0.2 m, and plot equally fine profiles from both. The potential consequences of using the less accurate set of data, however, are severe: the results may be totally useless or, at best, cause gross miscalculation of construction costs.

Whenever original data are based on digitisation of existing maps, the cartographic adaption may introduce error. For instance, roads are not defined on a map by their edges, but by solid lines. A line representing a railway may have been moved slightly so that both it and a second line representing a road, when both lie close to a coast, can be shown clearly. Maps of scale 1:50,000 are more generalised than maps of scale 1:5,000, and are as incompatible for GIS uses as they are for viewing together. Cartographic adaption can even mar the accuracy of a large-scale map, such as a detailed map of an electrical supply network, where the lines drawn may be moved in order to draw all the electrical cables connected to a substation.

The accuracy of georeferencing in digital data also depends on the densities of points used to represent objects, such as the inter-point interval selected in digitising. If the inter-point interval is relatively large and the lines digitised have sharp curvatures, accuracy will be less than if the inter-point interval was smaller.

Ordinary maps comprise the most common sources of digital map data. Hence, there are several successive sources of inaccuracies in digital map data, including original surveying, photogrammetric map making, map drawing and digitising.

These errors can be cumulative; see Eq. 7-3. If they are, they can have an adverse affect on the final data.

Digital map data generated directly from aerial photos or entered during surveying entail fewer steps and consequently are less subject to error than data from digitised maps. From the standpoint of accuracy, original sources are always preferable to maps.

Digital data storage precision can also affect data accuracy whenever entries comprise many significant figures. Storage with single precision, in which a single computer word (string of bits) is used to represent a number, results in precision of six or seven significant figures. Storage with double precision, in which double the usual number of bits are used to represent a number, results in precision of up to 14 significant figures. If data are transferred from a single precision system to a double precision system, digits after the 6th or 7th are unreliable. Single precision systems may also introduce noticeable error in rounding off, as when computing nodes.

All conversions from raster to vector presentations, and vice versa, entail some loss of accuracy.

All data should carry accuracy designations so that GIS users may assess the consequences of the various sources of error involved. Ideally, accuracy codes should be assigned to various objects to designate standard deviations and/or to attest source details, as map scale, aerial photograph scale, surveying parameters, photogrammetric registration, digitiser settings, etc.

7.4   ATTRIBUTE DATA ACCURACY

Considerations of accuracy usually focus on the accuracies of geometric data, although in practice the accuracies of attribute data are equally important.

Ill-defined attribute data can introduce errors in final data, for example when incomplete definitions of object types result in objects being wrongly classified. Thus, a residential building might be classified as a commercial building if the demarcation between the two classifications is unclear.

Quantitative data may be defined in non-numerical terms (ordinal data), as discrete variables divided into classes (interval data) or continuous variables without numerical limits (ratio data). Regarding accuracy, continuous variables may be treated as georeferenced data, see Eq. 7-1. For discrete data, the situation is different. If, for instance, discrete data classes are incompletely defined, the final data may contain errors arising from unintentional mis-classification. In some cases, mis-classification may involve simple sorting errors: an object of type A is put in class B. In other cases, the class structure may be faulty, as when there is no class C for objects containing elements of both A and B.

In practice, boundaries between areas of differing characteristics, say between different types of vegetation, are often inaccurate because no two persons interpreting the original data will draw exactly the same boundaries. Fragmented boundaries, as often occur in nature, cannot be measured accurately, regardless of the instruments used or the skill of the operators. [Which may be why fractals so befuddle GIS experts.] Operators encoding data may be unequally skilled and may therefore introduce systematic error, and whenever borders are ill-defined, the information of the polygons they delineate may be questionable.

Furthermore, measurement instruments may be poorly adjusted poorly, thus introducing systematic errors. Or they may be subject to random inaccuracies, as for instruments measuring acidity and alkalinity in pH, or for satellite sensors viewing reflected and re-emitted solar radiation.

The methods used to collect attributes also influence attribute accuracies. Examples include collections based on statistical distribution hypotheses or

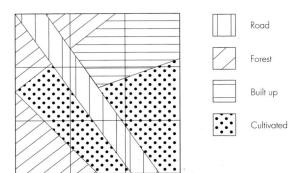

Road

Forest

Built up

Cultivated

Fig. 7.3 – The observation unit (cell size) selected here is too large to register all the objects desired.

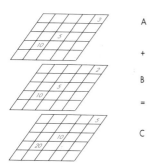

A

+

B

=

C

Fig. 7.4 – During addition of the attribute values in a raster layer, errors are transmitted to the total.

other assumptions that fail to agree with the real world, as when observations are excessively disperse. This is obvious in the choice of cell size in raster GIS, as shown in Fig. 7.3, where the density of observations (cell size) is too small to register the phenomena desired.

Coding errors rising when manual register data are entered in GIS may also diminish accuracy.

Generally speaking, the accuracy of numerical attributes is verified by comparing data with randomly assigned true values, and expressed in terms of standard deviations and systematic errors.

Errors can accumulate even as data are being processed. For instance, when two overlays of a raster GIS, A = 3 and B = 2, with standard deviations $m_A = \pm 1$ and $m_B = \pm 2$, are multiplied to a resultant C, the standard deviation m_C is as given in Eq. 7-4, provided that A and B are uncorrelated.

$$C = 3 + 2 = 5$$

$$m = \pm \sqrt{1^2 + 2^2} = \pm 2.2 \qquad (7-4)$$

The many probable sources of attribute data error can impose constraints that restrict the validity of data and limit the accuracy of final analyses.

A suitable expression of classified attribute data accuracy is to state a percentage of correct or erroneous classifications, such as "99% of all objects correctly classified."

Ideally, the accuracy of all attribute data should also be accuracy coded as geometric data. However, this is seldom done because the accuracy remains unknown, the cost of assessing it exceeds the benefit of the knowledge it provides, or because the user fails to appreciate the need for accuracy.

7.5   GEOMETRY-ATTRIBUTE LINK CONSISTENCY

Linking geometries and attributes correctly depends on the same i.d. codes being used in both sets of data. Consider, for instance, a situation in which road geometry data are stored in one database and road attribute data in another. The roads are divided into numbered segments, and the numbers serve as common i.d. codes. When new roads are built, new i.d. codes are added in the databases. The two databases should be updated simultaneously. If they are not, link inconsistencies may arise, diminishing the quality of the data jointly stored in the two databases. Such problems arise most readily whenever several agencies or users enter data into common databases at different times.

It is therefore advisable, whenever shape and attribute data are to be linked, to compile tables of missing links as data are entered. The consistency of linking is customarily stated in terms of a percentage, e.g. "99% successful links."

7.6   LOGICAL RELATIONSHIP CONSISTENCY

The logical relationships in data must satisfy the requirements imposed on relationships between objects which, in turn, are based on the tasks to be performed. Thus, polygons bordering on each other must have common borders; roads must connect in a logical network; polygons must be closed; depths must refer to the shoals involved, and so on.

The consistency of logical relationships is difficult to measure. In any case, quality depends on the use of a checking program as data are entered; data of good quality are used to correct data entered to form topologies.

Logical connections in map data are essential for GIS users, though less important for others. Users of map editing systems, for instance, may manage perfectly well with spaghetti data, but poor quality topology complicates or introduces error in many GIS functions, not least in overlayering and network analysis.

Consequently, data with unverified topology may cost a user dearly in terms of verification and correction work, or in terms of degraded quality of the final product.

7.7   DATA COVERAGE AND DATA RESOLUTION

Data completeness describes how completely data on an object type have been entered, for example whether attribute data as well as digital map data have been entered for all properties, all roads, all buildings, etc., in an area.

Completeness may be verified by comparing the number of objects for which data are entered against the number of real objects in the area involved. The result of the comparison may be expressed in terms of a percentage. In general, the more attributes that are entered, the better the description. Yet there are no objective measures that express how completely an object should be described. So verification is relative and can be expressed only in terms of a specification compiled for the purpose.

Resolution denotes the density of observations – the cell size in raster GIS, the size of a polygon in vector GIS, or priority of measurement points; see Fig. 7.3. The respective qualities of completeness and resolution are often inter-related. For instance, in Norway, all properties with an area of half a hectare or more are registered on 1:5,000 scale maps. Some areas on these maps may be completely covered, yet the coverage of all properties within the boundaries of the map may be incomplete. In this case, the observation density is not sufficient to depict all the properties completely.

Completeness is decisive for some objects. For instance, local or national regulations may require that neighbours be given notice whenever a property is divided. GIS may be used to locate those neighbours. Should just one neighbour be overlooked because of incomplete database completeness, the entire notification procedure may have to be repeated.

Whenever small-scale maps are used to generate digital map data, the generalisation imposed by the constraints of scale may degrade data coverage. So although the houses of a group may be shown individually on a large-scale map, a single symbol may represent them all on a small-scale map with less precise resolution.

Consequently, completeness and resolution should also be declared for data entered and stored, in order to inform users of the limitations entailed.

7.8   DATA CURRENTNESS

Two situations may affect the currentness of data:

1. When the geometries and attributes
   of existing objects have been altered.
2. When new objects, with geometries
   and attributes, have appeared.

The degree of currentness required depends on object type and data application. Public works agencies and utilities often require the most up-to-date data on property boundaries, buildings, roads, manholes and other works. It is therefore wise to update the data for such objects every two to three weeks.

Zoning boundaries are typical of the kind of data that change more slowly, and thus require less frequent updating. However, attributes may change more rapidly than geometries. Examples of this include new definitions, new registration methods and changes of vegetation during a growth season. Another common case is that of property data, where ownership changes more often than the boundaries.

Lack of current information on maps has always troubled users. Studies conducted in Norway have shown that the use of outdated 1:1,000 and 1:5,000 maps can cost each professional user an amount corresponding to one-third of the annual salary of a technical staff member. This is because map data must be supplemented, often with in-field measurements. Given the number of professional users in the country as a whole, the aggregate cost of outdated map information could clearly be enormous.

Currentness of data becomes particularly important whenever tasks involve data from various sources. Currentness is customarily indicated by the date of registry, such as the dates on aerial photographs, photogrammetric data and satellite data, and the dates of surveys, accidents, water leakages and the like.

7.9   PROBABLE SOURCES OF ERROR

Error is inevitable in all measurements, be they quantitative or qualitative, and they may range from the serious to the negligible. Yet neither error occurrence nor magnitude are known in advance.

As discussed above, errors and omissions degrade the accuracy of georeferencing, attribute data, logical consistency, linking, coverage and resolution, and currentness.

Error may be introduced in any phase of data processing. The following conditions (adopted from Burrough, 1986) may affect data quality.

A. Original error sources independent of GIS processing.
1. All surveys and field registrations. Errors may be ascribed to:
   - instrument inaccuracies
     - satellite sensor systems
     - aerial cameras
     - GPS
     - surveying instruments
   - various instruments for measuring attribute values

2. Data processing, as when producing maps used as GIS sources. Errors may be ascribed to:
   - map making
     - errors in computations and geodetic networks
     - mapping instrument inaccuracy
     - map drawing inaccuracies
   - data editing
     - computations
     - enlarging/reducing and redrawing

3. Errors due to changes in the field
   - registered objects change character
   - new phenomena arise

4. Errors introduced due to lack of coverage or resolution.

B. Errors introduced in GIS data processing

5. Errors in data entry
   - inaccuracy in digitisation
     - equipment error
     - operator error
   - inaccuracies in entering attribute data
     - human error (lack of verification routines)

6. Errors in storing data
   - insufficient computer numerical precision
   - storage medium errors

7. Errors in data manipulation
   - raster to vector
   - vector to raster
   - generalisation and thinning
   - combining classes
   - overlayering
   - interpolation (contours, elevations, etc.)
   - analyses of satellite and other data

8. Errors introduced in data presentation
   - plotter inaccuracy
   - paper and other presentation media inaccuracies

C. Errors in methods
   - errors in methods used to collect data
     - insufficient observation density
   - ill-defined objects and classes
   - insufficient expertise of data compilers
   - uncertain borders between areas

Improper Uses of Data

Data may be improperly used because they are not understood properly or because users lack the requisite expertise.

Even hypothetically error-free data may be improperly used.

Good data descriptions and specifications of accuracies, as well as user expertise in the relevant application are the best assurances of proper data use.

A few Rules

A few simple rules may help to reduce the problems of inaccuracies:

- Employ verification routines to ensure quality.
- Verify data as early as possible.
- Verify data at several stages of their manipulation.
- Don't mix data of high and low accuracies.
- Know the nature of the data, be it geometry data or attribute data.
- Be critical in all data uses.
- Apply processing results carefully.
- State inaccuracies associated with results and analyses.

8. DATABASES

8.1 INTRODUCTION

Traditional "libraries" and "banks"

Map data have traditionally been recorded in the form of lines and symbols on foil and paper. Similarly, descriptive data have been recorded in written form on file cards, in lists and various documents, and in various types of photographs. These traditional data repositories are organised in various systems of filing cabinets and drawers, loose-leaf binders, folders and other document files. Each data repository may be regarded as a "library" or "bank" from which users may retrieve information.

Conceptually, traditional information libraries and banks render data readily accessible because recorded data can be physically handled and viewed. However, as discussed in previous chapters, there are many drawbacks. For GIS purposes, the major disadvantages are:

- Data are dispersed. Data may be located in the files of several agencies or organisations, with no ready means for transferring the data from one storage facility to another.
- Dissimilar structure and storage. There are so many different ways of compiling and storing data that data from disparate sources can seldom be used jointly without extensive manual translation from one form to another.
- Verification is uncertain. There are no ready means of recording or authenticating data verification.
- Retrieval is slow. People must retrieve data: the physical search, and transport of sheets and documents consumes time.
- Data normally are available only to a few users.
- Data are restricted largely to the uses for which they were originally compiled; using such data for other purposes is difficult, if not impossible.

Data libraries and banks

Computerised data libraries and databanks differ from their traditional counterparts in a variety of ways, most noticeably in that all data are recorded in digital form and physically stored on magnetic or optical media. Initially, these electronic means of recording and storage may seem less accessible than their traditional counterparts because the recorded data cannot be handled physically and viewed.

The conventional repository for data is a databank, a computerised system for the "deposit" and "withdrawal" of data. A databank may either be avail-

Fig. 8.1 – The data which is fed into the database has to be structured in such a way as to produce useful applications and products.

able to a wide range of users or restricted to only a few authorised users. The data "deposited," or entered, may be in the form of one or more files, or in the form of a database. The difference between a file and a database is semantic and varies somewhat from one discipline to another. For GIS purposes, a file is regarded as a single collection of information that can be stored, whilst a collection of files is regarded as a library. A database comprises one or more files that are structured in a particular way by a Database Management System (DBMS), and accessed through it. Therefore, several databases can be created from a single file. Apart from such temporary working files as word processing text files, which may be entered in GIS, almost all computerised GIS information is stored in databases.

As discussed in preceding chapters, real-world phenomena are structured, i.e. grouped and described, in order that they may be understood more easily. Structured geographic phenomena are described in these phenomena in terms of geometries, attributes and relationships in a data model. A data model may be delineated digitally using coordinates for points, lines and areas, to which codes may be assigned.

The advantages of databases and Database Management Systems compared to traditional "libraries" and "banks" are that:

- Data are stored in one place.
- Data are structured and standardised.
- Data from dissimilar sources may be interconnected and used jointly.
- Data are amenable to verification.
- Data may be accessed rapidly.
- Data are available to many users.
- Data may be used directly in many different application programs, including programs whose purposes differ from those for which the original data were compiled.

The disadvantages of database storage compared to traditional storage are:

- Use of databases requires expertise.
- The products used are relatively expensive.
- Users may have to adapt to a data flow that differs from the traditional.
- Users may have to adapt to a data organisation that differs from the traditional.
- Data may easily be misused.
- Data may be easily lost, necessitating special security measures.

In other words, despite their many advantages, databases are no panacea. They can be misapplied and overused. The homeowner who uses a database on a PC in order to keep an updated record of the contents of the home freezer is solving the daily "what's for dinner?" problem with technological over-kill; a pencil and paper next to the freezer is more efficient.

Likewise, the choice of a DBMS ill-suited to an application may escalate both work and costs. Database structure is perhaps most important because it affects the response time, storage capacity, flexibility and other parameters which together determine the user friendliness of a database.

Structures differ and no one structure is superior for all applications. The various methods of data storage differ primarily according to the facilities or operations supported by the Database Management System:

Files (no database management):
- Tabulations of data in files. Databases, called flat-file databases, may comprise files only, with no supporting hierarchy or other inter-relationships.

Database Systems, distinguished according to the type of database management involved:
- Hierarchial database systems
- Network database systems
- Relational database systems

8.2   FILES

Files comprise records, each of which contain fields. Each record contains data concerning a single topic or affiliation; each field contains an item of data consisting of one or more characters, words, or codes that are processed together.

Keys, which are codes used to access information, help to retrieve records from files. Keys are associated with one or more fields of a record.

Data in a file may be stored sequentially, as on a line. New records are simply added at the end of the file. Records may be of variable length, in which case the beginning of each record must contain information on its length, although such information is not necessary when the length of the record is fixed.

	Field 1	Field 2	Field 3
Record 1	110	Oslo	10.61
Record 2	115	Stockholm	9.15
Record 3	116	London	18.33
–			
–			
Record n			

Fig. 8.2 – A file is divided up into records and fields.

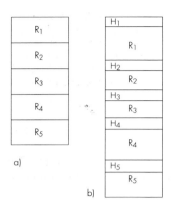

Fig. 8.3 – There are two categories of sequential files: a) fixed record length (R1, R2) and b) variable record length. The headers, H1, H2 etc. specify the length of the record.

Searching for data in sequential files is time consuming. For instance, if the search is for points stored in an unordered sequential file of records of points specified by their x and y coordinates, then all records in the file must be queried to ascertain whether the desired points have been located. Therefore, storage in unsorted sequences is recommended only for relatively small quantities of data.

The search time may be reduced if the data in a file are structured, for instance, when they are sorted by ascending coordinate magnitudes. The search may then be limited by the largest magnitudes entered. A search may start in the middle of a file. Therefore, searching may progress successively, by dividing intervals into two, deciding each time which half contains the point sought, and carrying on until the point is located. This process is known as a binary search and is considerably faster than sequential searching.

The address, or position of an item in a file may be stored in another file. Known as an index file or pointer file, it contains a set of links that can be used to locate records in the data file. Searching for items in the data file may be compared to searching for entries in a well-indexed book, by first finding page numbers in its index.

8.3   DATABASE SYSTEMS

Unstructured files are ill suited for large quantities of data, or for data that change continually. So, as multiple database storage may be common and amending data complex, Database Management Systems (DBMS) have been developed to overcome these shortcomings. The data in a database are stored on disks or tapes, so a database is a physical collection of files structured by a DBMS.

All database systems aim for ease of search and linking of tabular data. A Database Management System must be able to manipulate several types of objects and various relations between objects.

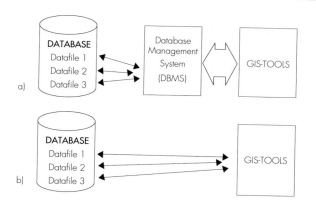

8.3.1   HIERARCHIAL DATABASE SYSTEMS

A hierarchial database system is one in which the DBMS supports a hierarchial structure of records organised in files at various logical levels with connections between the levels. A record at a particular level contains data common to a set of records at the next lower level. There are no connections between records of the same level. Each record contains a field defined as the key field, which organise the hierarchy.

Records of the same type are collected in files known as elements. Several types of elements may reside at the same logical level. Starting from the high-est-order element, the hierarchy permits a set of elements to be accessed at the next lower level. Any one element may retain connections to only one set of lower-level elements and itself may be a member of only one such set.

Construction of a hierarchial structure begins with a main object at its top. The main object has a range of characteristics which can be collected at the various levels of the hierarchy.

Geographic data may be stored in a hierarchial structure in a manner that reflects the real world. An example is when the levels of a hierarchial model correspond to the levels of real-world administration, for example in a country, county, township or town/city.

Hierarchial database systems are easily expanded and updated. However, they require large index files, must be frequently maintained and are susceptible to multiple entries. Searches are rapid, but search routines are fixed and constrained by the structures. Thus, records at the same level cannot be searched, nor can new links or new search routines be defined. The elements or the structure are related only through one-to-many connections.

This constraint imposes the presupposition that all queries are known in advance and accounted for in structuring and entering data. This constraint is not always natural or suitable in GIS applications. As a result, hierarchial database systems are usually restricted to storing digital map data in GIS.

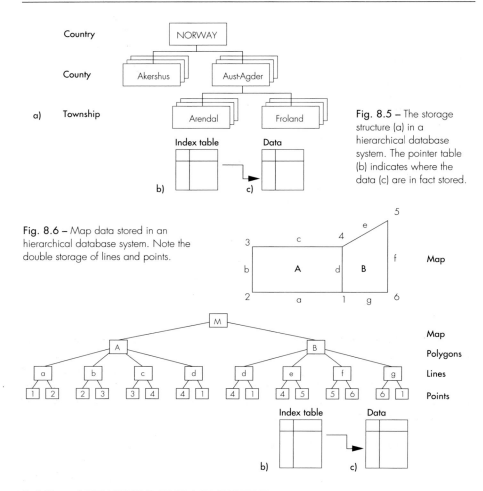

Fig. 8.5 – The storage structure (a) in a hierarchical database system. The pointer table (b) indicates where the data (c) are in fact stored.

Fig. 8.6 – Map data stored in an hierarchical database system. Note the double storage of lines and points.

8.3.2   NETWORK DATABASE SYSTEMS

A network database system is one in which the DBMS supports a network type of organisation. Each element, or collection of like records, has connections to several different-level elements. Interconnections are made in the hierarchial organisation, and a characteristic may be associated with two main objects. The resultant network structure more closely represents the complex interrelationships which often exist between real-world geographical objects. The elements of the structure may be related through one:many, many:one and many:many connections.

The purpose of the network structure is to improve the flexibility and reduce the multiple entries of the hierarchial structure. Points and lines are entered only once. Searches need not pass through all levels, but can take short cuts. However, the volume of index data is greater than that of the hierarchial structure.

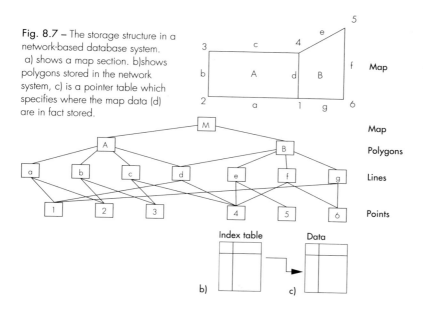

Fig. 8.7 – The storage structure in a network-based database system. a) shows a map section. b)shows polygons stored in the network system, c) is a pointer table which specifies where the map data (d) are in fact stored.

A network structure permits rapid connection between data which physically are stored in different disk sectors. However, maintaining data stored in a network structure is complex, so although a network structure is better suited to geographic data than is a hierarchial structure, it is infrequently used in GIS applications. However, network structures are frequently used in administrative data systems.

8.3.3   RELATIONAL DATABASE SYSTEMS

A relational database system is one in which the DBMS supports a relational model of the data stored. That is to say, it is more of a guideline to representing data than it is a structure for data. In many ways, it is comparable with a table, which is why "tabular database" is an alternative term for relational database.

In the relational database system, there is no hierarchy and no specific keys point to the various records. All fields in a line may be regarded as being permanently connected to each other, yet individual tables contain no information on other tables. Nevertheless, any field in a table may be a key for accessing data from another table. The system permits all objects and attributes to be related to each other. This means that a collection of data on many objects with complex relationships may result in a large number of tables – a weakness of the relational approach.

There are no pointers in the data tables. Internal index tables are used to direct inter-table communications, which means that the system must open at

Building I.D.	Property	Owner	Year	Type
589	44 110			
610	44 50			
955	44 99			

Property I.D.	Owner	AREA	ADDRESS
44 99			
44 50			
44 110	Nils Nilsen	6,51	9999 Toppen

Fig. 8.8 – A relations database. Each field in a table can be the key to locating data in another table.

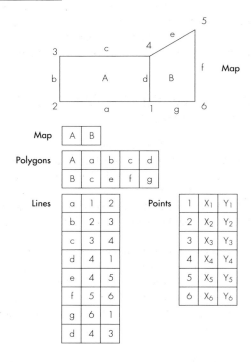

Fig. 8.9 – Examples of map data stored in a relations database.

least one index table for each connection between data tables. This can result in relatively large databases and slow access. Moreover, as there are no pointers, searches for records or fields must be sequential through the tables. The result is that relational database systems are inherently slower than hierarchial or network data base systems.

However, a field may be located quickly in a given table, and the data accessed may be manipulated. The simple structures of relational database systems have permitted the development of standard query languages, one of which is Standard Query Language (SQL).

Relational algebra may be performed using two classes of storage and retrieval operations. The set operations include union, intersection, difference and product. The relational operations include selection (accessing rows), projection (accessing columns), joining and dividing. Joining creates a new table from data retrieved from various tables. The new table need not physically be stored in the database.

Relational DBMS are now the type used most frequently in GIS applications, primarily because of their simple, flexible structures. Another reason is that they support the complex relationships common among real-world geographic objects. Multiple entries are less frequent than in hierarchial or network database systems, although search times tend to be longer. This is particularly noticeable in specific operations on digital map data, for instance when forming polygons and with automatic text placement.

8.4   ADAPTING DATABASES TO GIS USES
8.4.1   GENERAL

Within the framework of GIS, data are logically divided into two categories: geometric data and attribute data. This division may also extend to a similar division of physical storage, although the relationships between the two categories of data must be preserved, regardless of whether the division is logical or physical. In current GIS, this condition is satisfied through one of three approaches to the overall database storage involved:

- a single database system that stores both geometrical data and attribute data
- two separate database systems, one for geometrical data and one for attribute data
- one database for geometrical data connected to several different databases for attribute data

Systems that store geometrical data and attribute data in two separate databases are usually termed hybrid systems.

There are also several approaches to Database Management System tools. The DBMS tools for databases storing both geometrical data and attribute data tend to be specialised. Systems with separate databases normally employ DBMS that are commercially-available for attribute data, and tailor-made DBMS for geometrical data.

Most GIS users have developed application programs with various man-machine interfaces. Standard Query Language (SQL) is used most frequently in searching, although other query procedures are also used.

Systems that combine geometrical data and attribute data in a single database most often employ a relational database with a geometry of nodes and links that is structured topologically. The data structure is usually object oriented.

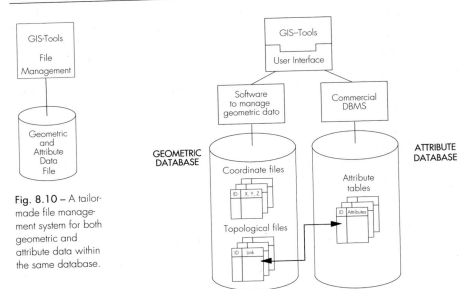

Fig. 8.10 – A tailor-made file management system for both geometric and attribute data within the same database.

Fig. 8.11 – The hybrid database model, where geometric and attribute data are stored in separate databases.

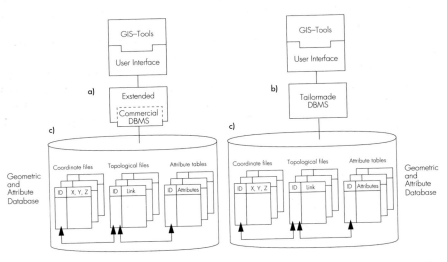

Fig. 8.12 – The integrated GIS model, with two alternative solutions for database management (a,b) and with detailes of the Integrated Database (c).

Hierarchial databases, network databases or combinations of network and relational databases are used whenever the data to be stored in a database are restricted to geometrical data alone. Relational or network databases are employed whenever attribute data alone are stored.

The current trend in GIS follows the general information systems propensity towards general, movable programs and tools.

Some systems feature special functions dedicated to geometrical data, which enhance yields in manipulation and presentation. Such special functions, however, impose constraints that may complicate their use with general programs. Systems that employ a single database or commercially-available DBMS tools are often more easily adapted to work with newer database systems.

Rapid advances in technological development may, in the future, make the storage of common data in a a physical shared database unnecessary. Storage can be divided between several producers/users. The manipulation of common data (search, selection, transport and organisation) is carried out by file servers in a data network. GIS will manage a logical database, which physically could be stored in either one or several different locations.

8.4.2   DATABASES FOR MAP DATA

Compared to network databases, relational databases for digital map data have some drawbacks and limitations which restrict their applicability in specific searches and prolong their response time. Thus, in editing and updating, the system must continuously update all relations and pointers, which consumes time.

Databases for digital map data should be able to manipulate records of varying length efficiently. For instance, line length may vary considerably, resulting in a corresponding variation in the number of coordinates entered. Standard databases cope poorly with such variations. This is why many GIS suppliers have developed their own databases for digital map data. These database systems reflect geographic reality, for instance by requiring that data

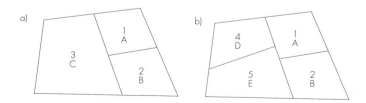

Fig. 8.13 – Updating is complex.
Property 3 has been divided and has two new owners, D and E. Property 3 has to be deleted from the database and the two new properties, 4 and 5, registered. Property 2 now has property 5 instead of property 3 as its neighbour. Property 1 has lost one neighbour (3) and gained two new ones (4 and 5).

on objects of the same type, such as the lines forming a property boundary, be stored in close proximity in the database in order to speed up the response. This feature is not possible in hierarchial or network systems which lack linking between objects of the same group.

Tailor-made database systems using relatively simple file structures are often used for geometrical data. These databases are usually based more or less on network or relational databases, or a combination thereof.

Map databases for vector data store:
- point counts
- codes (i.d. codes etc.)
- coordinates
- plotting routines (how outputs are drawn)
- search data (pointers)

Distributed Storage of Vector Data

Generally, map data are stored in map sheets or other geographic units, but storing map sheet data in single sequential files lengthens the response time. This has resulted in some GIS employing addressing to speed up the searching process, and enabling current map sheets to appear on screen almost immediately. Addressing specifies locations, so map sheets are divided into sections which are distributed in a such a manner as to accelerate the search. For instance, zooming focuses on data in those sections relevant to a selected area and ignores the remainder of the map sheet.

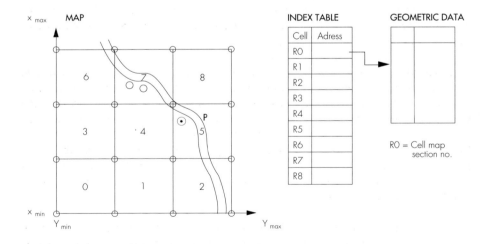

Fig. 8.14 – The manipulation of digital map data is made more efficient by storage in small squares according to a special addressing principle. The pointer table enables the system to access the data quickly once it has established in which square the data are located.

The underlying concept is that a key is used to locate the data physically on a disk (Bjørke, 1989). Each map section is assigned a number, which is listed in a table of pointers to storage locations. The map section numbers, and hence the entries in the pointer table, may be calculated from Cartesian coordinates.

The time required to bring data on screen increases as the amount of information in the map sections grows. In other words, the smaller the sections, the more rapid the search. But smaller sections mean more pointers, which slows retrieval – so compromises are often necessary.

Hierarchial Data Structures for Raster Data

As discussed in § 4.3.2 of Chapter 4, raster data may be stored in compressed form using run-length encoding. Raster data may be even more efficiently stored in quad-tree hierarchial database systems. The quad-tree paradigm divides a geographic area into square cells of size varying from relatively large down to that of the smallest cell of the raster. Usually, the squares are successively quartered into four smaller squares.

The quartering may be continued to a suitable level, perhaps until a square is found to be homogeneous, so that it no longer needs to be divided, and the data on it can be stored as a unit. A larger square may therefore comprise several raster cells having the same attribute values. However, homogeneous areas that are not square or do not coincide with the pattern of squares employed, may be further divided into homogeneous squares.
Quad-tree storage requires two tables: one for attribute data for homogeneous areas, and one for pointers that locate the attributes of each area in the data tables.

Each square may be assigned a number according to the Morton Index, which, as shown in Fig. 8.19, shows a successive number of squares as they are quartered. Numbers of successive quarterings are combined, so that the index number indicates the size of the square (a low number indicates a large square), as well as the relative geographic location and hierarchial location in the pointer table. Larger areas will have pointers located higher up in the pointer table, so that their data may be retrieved from the attribute table rapidly. The attributes of neighbouring squares on the ground are stored as neighbours in the attribute table.

As shown in Fig. 8.18, the structure of the pointer table resembles an inverted tree, whose leaves are the pointers to the attributes of homogeneous squares and whose branch forks are pointers to smaller squares. Hence the name quad-tree.

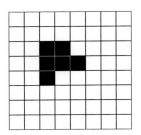

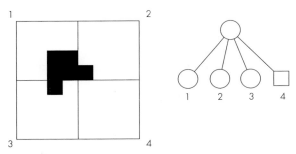

Fig. 8.15 – Quad-tree raster encoding recursively sub-divides a map into quarters to define the boundary of an area. The shaded squares on this grid represent the region which is to be stored in a quad-tree. (adapted form Green & Rapek)

Fig. 8.16 – Initially, the grid is divided into four quadrants. If part of the region falls within a quadrant, it is assigned a white circle. If no part of the region falls in a quadrant, it is assigned a white square. (adapted form Green & Rapek)

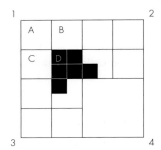

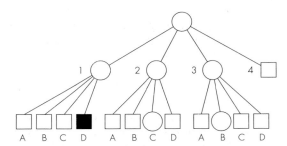

Fig. 8.17 – The quadrants which contains part of the region are then sub-divided again, which adds another level to the tree. If a quadrant is completely filled by part of the region, it is represented by a black square. (adapted form Green & Rapek)

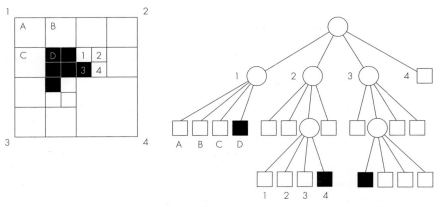

Fig. 8.18 – Sub-division of the grid continues until each "leaf" in the tree is represented either by a black square or a white square. Only 19 leaves are required to store this region in a quad-tree. (adapted form Green & Rapek)

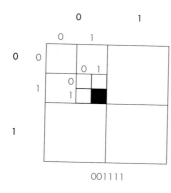

001111

Fig. 8.19 – An example of a Morton index with binary numbered cells. Black = 001111

The advantages of quad-tree storage are:

- rapid data manipulation because homogeneous areas are not divided into the smallest cells used
- rapid search because larger homogeneous areas are located higher up in the pointer structure
- compact storage: homogeneous squares are stored as units
- efficient storage structure for certain operations, including searching for neighbouring squares or for a square containing a specific point

The disadvantages of quad-tree storage are:

- establishing the structure requires considerable processing time
- protracted processing may prolong changing and updating
- data entered must be relatively homogeneous
- complex data may require more storage capacity than ordinary raster storage.

8.4.3   OBJECT ORIENTED DATABASE SYSTEMS

Hierarchial, network and relational database systems are essentially intended for administrative tasks. Consequently, they are not particularly well suited to representing a conceptual data model of geographic reality. All these systems, however, are record oriented, which is to say that the data they contain are filed record by record, rather like cards in a conventional card file. For instance, in a relational database, all the elements which together comprise a particular map object may reside in several records in various tables. Furthermore, geometrical data and attribute data are often physically separated, each residing in a separate database. Topological storage of vector data is object oriented, but the relevant database structure remains a problem.

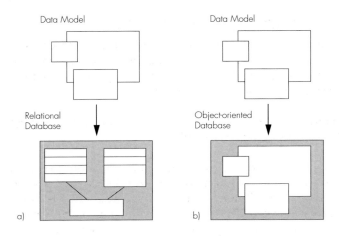

Fig. 8.20
– Object-oriented database (b) compared to a relational database (a). (adapted from Jan H. Lindblad)

Object-oriented database systems attempt to overcome these difficulties by representing more faithfully the real world, which comprises homogeneous, though complex, objects with varying internal and external relations.

In object-oriented systems, data modelling and data manipulation are in many ways synonymous. Objects are manipulated as independent homogeneous entities, yet may comprise several other objects in a hierarchy.

Object-oriented database systems employ simple, albeit "intelligent" databases. However, the requirements for user expertise are correspondingly higher because users must be able to define all relationships between objects in the database system.

The chief advantage of object-oriented database systems is that they are easier to update on a consistent basis. For instance, a line (which is an object) may be a property border, a road edge and a zoning boundary. In an object-oriented database, it need be updated only once, whilst in other database systems the data pertaining to it may be in various locations, requiring multiple updating operations.

To date, object-oriented database systems are not widely used in GIS, although a few GIS use them in manipulating geometrical data.

Conclusions

The usefulness of various database systems may be evaluated in terms of their applications.

No current database system completely fulfils the needs of database applications. There are grounds for suspecting that the excessively complex and voluminous data collections of many GIS may be ascribed to the databases employed. It goes without saying, then, that further database development is in order. One goal might be to develop better object-oriented database systems.

8.5 SAFEKEEPING AND SECURITY ROUTINES

In many ways, computerised information is more vulnerable than conventional information. Errors in equipment, storage media and computer programs, as well as user mistakes, can have serious consequences. Therefore, security routines are necessary to guard against errors.

With the collection of geographic information so costly, damage to storage media (tapes, disks etc.) can result in considerable loss.

Moreover, as updating, access and new data entry become more frequent, security needs increase. On-line users can hardly be expected to wait several days while data is reconstructed following damage or an accident. Electrical outages or losses of data are expensive and time consuming for both users and operating organisations alike. Security routines are therefore vital.

There are many causes of data loss. The storage medium may be damaged through misuse. Equipment may fail. Fire or other damage may occur. In daily use, data in a primary memory may be lost due to user mistakes, operative system errors, equipment failures or electric power failures. Without prior warning, disk sectors or entire disks may become unreadable or data stored on tapes may deteriorate.

Backup copies of disk files should be made periodically in order to minimise reconstruction work should file data be lost. A customary rule for systems in daily use is that all database changes are automatically backed up daily from working disks onto tapes or tape cassettes. The entire database may be backed up less frequently, perhaps once a week or every other week. Some backup systems continually back up all transactions.

Working databases independent of original databases are often used in GIS work, for instance, for tasks concerning a limited geographic area. A working database should never be stored on the same disk as an original database, lest a total disk failure, or "disk crash" in computer jargon, should destroy both databases.

Data on tapes should also be backed up frequently. Although tapes store data well, data can in time deteriorate on all magnetic media. Therefore tapes should be periodically recopied; once a year is recommended.

Backup copies should not be stored with original data or databases, lest fire destroy both. Therefore, backup copies should be stored in a fireproof enclosure.

Some GIS contain data that must be guarded against unauthorised use, such as personal data, details of telecommunications networks and the like. The operative system can be configured to permit only authorised access to data or parts of data stored.

Complete safeguarding and security would be extremely expensive and time consuming. For instance, meticulous storage and handling routines can reduce the risk of disk damage, at a cost proportional to the measures enacted, and the disk storage boxes and cabinets acquired for the purpose. Uninterruptable Power Supplies (UPS) may be installed to guard against mains outages. However, these add to the equipment and may be costly – not least because each UPS unit consists of a rectifier that converts the mains AC into DC for float charging batteries, and an inverter for conversion from DC to mains voltage and frequency to power the equipment. An UPS to power a PC could cost more than half as much as the PC itself.

Therefore, safekeeping and security routines must be evaluated in terms of the costs of data loss. As in other sectors and disciplines, calculated risk must be taken into account. Insurance policies may be taken out against financial loss.

Normally, equipment may be insured against physical damage due to fire, flood, vandalism and theft. Some insurance is available to cover the costs of reconstructing data and the losses incurred during operational interruptions.

Finally, security measures may be required to guard against the potential losses of deliberate break in and data destruction, or data crime as it is now known.

9.   GIS TOOLS

9.1   INTRODUCTION

From the user viewpoint, the ideal GIS should include enough functions to perform all conceivable manipulations of geographic data. In practice, user needs comprise various tasks. So, as listed below, an overview of user tasks defines the overall GIS requirement (J. M. Larsen).

USER NEED (TASK)	DATA MANIPULATION REQUIRED
Catalogue access	Locate relevant data.
Requisition	Send data requested.
Entry	Enter and/or register selected data in the user system.
View	Study and evaluate received or registered data.
Select	Choose data required.
Correction	Rectify errors and omissions.
Compilation	Compile dissimilar data to a model or image of the real world.
Overview	Use model or image to provide a general overview of an area.
Monitor	Use model or image to monitor occurrences in an area.
Navigate	Use model to locate position of ship or vehicle and to determine quickest route between designated points.
Analyse	Evaluate connections, possibilities, conflicts and consequences.
Decision making	Ensure that data and data processing support choices of alternative actions.
Presentation	Present overviews, analyses and decisions in graphic form (maps and/or reports).

The first two tasks are customarily associated with GIS and hence are often incorporated into infrastructures under the jurisdiction of national map agencies, military agencies or public utilites. The remaining tasks are supported to a greater or lesser degree by the various GIS now on the market.

As a software system, a GIS usually comprises a set of software tools including:

- database
- database management system (DBMS)
- query language (QL)
- application functions and programs
- user interface

Databases, database management systems and user interfaces have been discussed previously, whilst query language has been mentioned in passing. The application functions employed in GIS are discussed in this chapter. These often employ standard functions supported by the database management system, the query language and the user interface.

The wide range of needs encountered requires numerous special GIS functions. Only a few GIS have all conceivable functions. In all cases, each function may be executed by one or more commands.

Special GIS functions may be divided into four main categories:

A. Functions for storing, registering and entering data
   • entry (including format transformation and similar operations)
   • organisation of storage operations
   • registration and verification

B. Functions for correcting and adapting data for further use
   • geometrical manipulation
     – editing
     – edge matching
     – map projection transformations
     – coordinate system transformations
   • attribute editing

C. Functions for processing and analysing data
   • query and report extraction from attribute data
   • processing attribute data
   • integrated processing of geometry and attributes

   Processing and analyses comprise:
   • measurements (of areas and peripheries)
   • altering attribute values
   • deleting and compiling objects
   • line thinning and smoothing
   • statistical computations
   • topological overlaying
   • network analysis
   • terrain surface operations

D. Functions for Data Presentation
   • use of symbols (cartographic variables)
   • text insertion
   • perspectives and other drawings

TOOL KIT

A: Registrering, entering and storage
B: Correcting and adapting
C: Processing and analysing
D: Presentation

Fig. 9.1 – The GIS software tool has four main components.

The four main categories above are based primarily on the various phases of most GIS projects, but in practice the categories may be mixed. For instance, functions for editing geometries may be employed both in editing and in presenting data.

9.2   ORGANISATION OF DATA STORAGE OPERATIONS

Software systems often organise data to ensure its effective use. Such organisation may involve various logical paradigms concerning the grouping of object types and the divisions of geographical areas. The physical limitations of system file capacities may also be a practical reason for thematic and geographic divisions.

Thematic Layers

Data in most GIS are organised in layers or on levels, much like the overlays of conventional map making. Similarly, individual data layers/levels are stored in individual data files. These layers may contain object types intended to be processed together, such as points in one layer, lines in another and polygons in a third layer. Alternatively, the individual data layers may be organized by theme, perhaps one layer for topography, another for property boundaries, and others for roads, types of land use and so on. Furthermore, each layer may contain subsidiary layers in a hierarchy. Thus, a layer for roads might also encompass separate subsidiary layers for national, county, urban and private roads.

Collecting logically similar objects can reduce the amount of data required to describe each object. Objects which represent several themes, such as lines

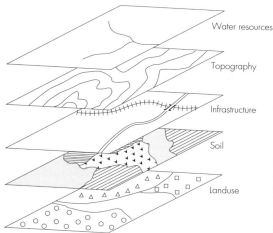

Water resources

Topography

Infrastructure

Soil

Landuse

Fig. 9.2 – An example of how data are organised.
Many systems organise data into different layers (files). Their efficient use is often dependent upon a well-thought-out organisation of data into a layer system.

which simultaneously are roads, property or land use area boundaries may be collected in one layer. The line geometry of that layer may then be transferred to other layers as needed.

Objects updated frequently or from the same source of information may also be collected in a single layer to facilitate updating work.

The cartographic effects of plotting are frequently dependent on the sequential plotting of layers containing like objects.

The separation of data into layers may seem analogous to the traditional separation of map information into overlays, and therefore not always a realistic data model of reality. Some newer GIS circumvent such "map overlay thinking" by being object oriented. That is, each object is manipulated as an independent entity, both with regard to its geometry and its attributes.

Partitioning the Area

All GIS have facilities to divide surfaces in order to promote efficient storage, use and the updating of data. The individual surface segments are then stored in individual files, division by map sheets being the most common. Some GIS support divisions of data structures into projects, each of which then may be further divided into sub-projects.

The manipulation of data for a larger area often involves combining data for its constituent segments. This is done either manually by the user or automatically by the system. As many GIS are seamless – i.e. data need not be regarded as belonging to fixed map sheets – users are presented with data representing contiguous surfaces, even though stored data may be divided into map sheets which, in turn, may be divided into cells in grids.

Users must choose the most suitable elements for storing data, such as various map sheet sizes and area divisions. Choice is vital for two reasons. Firstly, the organisation of data storage elements can influence considerably the efficiency with which data are used. Secondly, once the storage elements are chosen and data are accordingly stored, restructuring to other storage elements is extremely complicated.

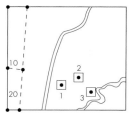

BR 012-5-2	BS 012-5-1	BS 012-5-2
BR 012-5-4	BS 012-5-3	BS 012-5-4
BR 011-5-2	BS 011-5-2	BS 011-5-2

Fig. 9.3 – An example of how data are organised into separate map sheets.
Many systems organise data into certain surface units, such as map sheets. Careful consideration has to be given to which storage element is most suitable, since restructuring to other elements is difficult once the data have been organised in a particular way.

9.3   DATA ENTRY FUNCTIONS

A GIS may be implemented with software supporting various forms of data entry, including entry from:

- digitisers
- photogrammetric instruments
- surveying stations
- GPS
- scanning and pattern recognition
- miscellaneous entry programs

The relevant programs include facilities for expressing original data in ground coordinates, for entering codes and for performing accuracy checks.

The software systems also have flexible modules for entering assorted types of attribute data, including the presentation of various screen displays.

9.4   IMPORTING EXISTING DIGITAL DATA

All GIS contain software that permits the import of existing digital data. The relevant processes entail transforming various exported formats to the GIS imported file format. Many GIS programs can read national or international standard formats such as ASCII, ISO, SIF, DXF, HPGL and so on, and some systems have routines for reading both vector and raster data.

9.5   FUNCTIONS FOR CORRECTING AND ADAPTING GEOMETRIC DATA FOR FURTHER USE

Data are ready for use only after they have been verified and, if necessary, corrected. All GIS contain programs for adapting data. These include:

- general utility functions
- an ability to create topology
- map projection transformations
- transformations to a common coordinate system
- adjustments at map edges or between other storage elements
- coordinate thinning and line smoothing

9.5.1   GENERAL UTILITY FUNCTIONS

Like all computer systems, GIS have collections of programs that provide such useful functions as file manipulation, editing and program cross-referencing. GIS incorporate other utility functions specific to GIS tasks.

Zoom, a utility for changing the scale of displayed screen images, is an example. Zoom may be activated either in the main GIS software or in the screen software. Screen zoom, which allows either direct enlargement or reduction, is the more rapid of the two. Although slower, GIS program zoom can include various intelligent functions that alter texts, line widths and other features in proportions differing from those of the overall scale magnification or reduction. In most systems there is a limit to how much an image may be enlarged.

Another GIS utility can highlight selected themes or objects, thus easing on-screen work with complex maps comprising many elements.

9.5.2   EDITING AND CORRECTING ERRORS AND OMISSIONS

Most GIS support geometric map editing functions including:

- supplementing
- copying
- deleting
- moving
- rotating
- dividing lines
- joining lines
- altering form

Raw digitised data always contain some errors and omissions. Some examples are lines crossing erroneously, unintentional wiggles in lines, missing points and lines, etc.

Omissions are most easily corrected by entering data directly from a digitiser. Errors are corrected most efficiently using a keyboard or a mouse.

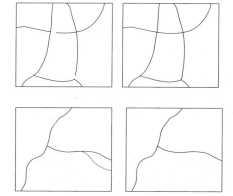

Fig. 9.4 – Polygons which are not closed can be corrected by automatically joining lines where the gap is below a given parameter. It should be noted, however, that automatic closing can have undesirable consequences, should the closing parameter be set too high.

Fig. 9.5 – Digital map data can be edited by, for example, the addition of new data and removal of old data, so that objects retain their correct shape.

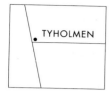
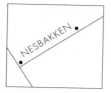
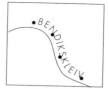

Fig. 9.6 – Normally, text can be freely placed with regard to start and end, rotation, following lines etc. Font types are generally system defined.

Lines may be divided or line segments deleted to allow texts or symbols to be placed on maps. Lines and composite map objects may be moved whenever objects are incorrectly digitised. Line segments to be moved or deleted are marked, as well as any new positions. In some cases, objects are rotated through prescribed angles.

Other utilities support text editing for altering or amending place names, road names and other names. Many systems support the direct printout, for example, of such attribute values as addresses and property registry numbers on maps. Some systems treat all texts as attributes.

Raster-based systems have programs with editors dedicated to editing raster data. Hence, systems that can manipulate both raster data and vector data have separate raster and vector editors.

The coordinate values may be modified using an ordinary text editor.

9.5.3   CREATING TOPOLOGY

Corrected data may result in an unacceptable map image, but even when such data is verified, the image may contain errors that are not evident until logical connections are established. Consequently, some GIS have functions that automatically compute nodes and links, and compile topology tables. Creating topology identifies such errors as:

- polygons that are not closed
- disconnected lines (dangling ends)
- missing or repeated i.d. codes
- polygons lacking, or having too many, i.d. points (in some systems, all polygons are assigned i.d. points with codes and coordinates)

Errors such as these may be reported by a dedicated error reporter or marked with specific symbols on plots. Errors may be corrected manually or automatically. Errors in i.d. codes and i.d. points are corrected manually via a keyboard.

Small, meaningless dangling ends can be deleted automatically by setting a parameter that stipulates the sizes of the ends to be deleted. Gaps may also

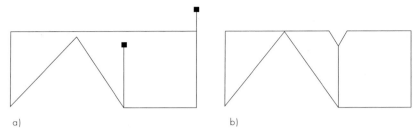

Fig. 9.7 – a) Various types of data error can often be identified during topology creation. b) Automatic error correction can be risky since it can introduce completely new, and highly undesirable, errors.

be closed automatically. However, if the gap closing parameter is set too high, automatic gap closing may introduce errors by connecting lines that ought not to be connected. Therefore, it is always advantageous to digitise line intersections in order to reaffirm line connections.

9.5.4   TRANSFORMATIONS TO A COMMON MAP PROJECTION AND A COMMON COORDINATE SYSTEM

Transformations to a common Map Projection

Data from a variety of sources are useful only if referenced to a common map projection. However, data from differing sources often are referenced to differing map projections. One common disparity is that survey points are entered in coordinates of latitude and longitude in an azimuth projection, whilst other map data are entered with reference to a national rectangular coordinate system based on a cylindrical projection. Consequently, most GIS support various transformations for converting data from one projection to another.

Mapping hemispheres and other major land areas usually entails selection of a map projection that most closely suits the data presented. So, in some cases, entire sets of data must be transformed.

These transformations are based on the mathematical relationships that describe the various map projections. Hence, a transformation converts the coordinates of one system to the coordinates of another. As the transformations themselves are digital, ancillary operations are necessary whenever positional information is not decimal, for instance when the coordinates are in degrees, minutes and seconds.

Transformations to a common Coordinate System

Data from various sources, as from map digitising and field surveying, may be used together only if referenced to a common coordinate system. Sometimes, systematic errors in such themes as displacements, rotations, scale

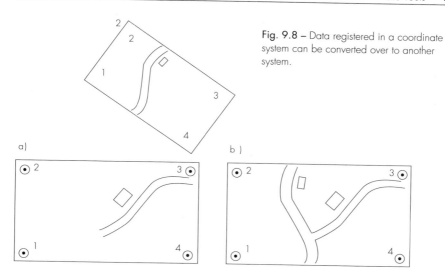

Fig. 9.8 – Data registered in a coordinate system can be converted over to another system.

errors can easily be compensated if the data involved are transformed to an "error-free" base. In some cases, the data of a particular thematic layer must be transformed in order to be compatible with the data of other thematic layers. Consequently, most GIS support a variety of coordinate system transformations.

Coordinate transformations involve mathematical functions that relate coordinate geometries to each other. A conversion function also contains parameters based on a knowledge of the respective coordinates of a number of points that are common to the initial and final systems of the transformation involved.

9.5.5   ADJUSTMENTS BETWEEN MAP EDGES AND BETWEEN NEIGHBOURING AREAS

As anyone who has glued neighbouring map sheets together to make a larger map knows, lines that should meet at the joined map edges often do not. Printing errors and non-uniform paper shrinkage are often the causes of this common vexing problem.

The problem is compounded whenever map sheets are digitised because digitisation may also introduce error. Therefore, most GIS support functions for automatic adjustment of disparities along neighbouring map sheet edges. This is hardly a trivial process because for each disparity along the meeting edges, a decision must be made to move one or both lines. When the process is completed for two adjoining edges, the polygon representing the edge may be deleted.

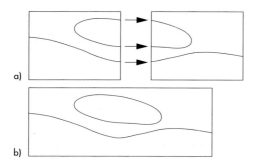

a)

b)

Fig. 9.9 – Map sheets can be automatically aligned and combined into one new seamless unit.

9.5.6   THINNING COORDINATES AND SMOOTHING LINES

Thinning Coordinates

Digitisation often produces more data than needed for the task involved, particularly with line objects, such as contour lines, borders for types of soil and vegetation, and so on. Consequently, most GIS have routines for thinning or deleting superfluous data. Thinning increases data efficiency, reduces the storage capacity required and increases the plotting speed. In some cases, data volume may be reduced by more than 60% (and up to 95%) without reducing the data quality.

Various methods are available for thinning points on lines. The simplest approach is to delete every nth point – every other point, every 3rd point and so on – up to the number n chosen. A more advanced approach entails moving a small, variable-width corridor along the lines. The number of points needed to describe a line may be reduced by moving forward a corridor of given width until it touches the digitised line. All the points on the line that lie within the corridor, apart from the first and the last, are thereby deleted. This process is prepeated until the whole line has been trimmed. Lines may also be replaced by polynomial spline functions, see Fig. 4.14 and 4.15.

Line Smoothing

Lines delineated by points are never completely smooth, so line smoothing is often used to improve the graphic quality of lines. Smoothing functions are usually associated only with data presentation.

Line smoothing means increasing the total number of points delineating a line. Several smoothing methods are available, but those used most frequently are based on curve fitting, using third-degree polynomials, as illustrated in Fig. 9.11.

Smoothing affects appearance only. Although a smoothed line may seem more accurate, it is no more accurate than the original data for the line.

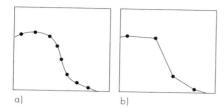

Fig. 9.10 – Lines can be thinned by removing points based on certain functions.

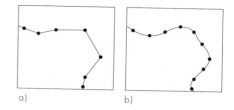

Fig. 9.11 – Height contours and other linear elements can be smoothed out by automatic addition of new points, based on certain functions.

9.6   EDITING ATTRIBUTE DATA

Like digital map data, attribute data must be edited and corrected. These operations include error correction, updating when data are changed and amending with new data.

The editing tasks may be carried out by using standard editing tools, as available in word processing, or specific GIS commands. Some GIS use SQL (Standard Query Language) to manipulate attribute data in relational databases.

Specific GIS commands include commands for changing object thematic codes and switching codes between objects, as well as for editing thematic codes containing texts.

The guidelines for entering data may change with time, mandating changes in the codes of older data. Usually, common mathematical signs, as +, -, * and / are used for this purpose. The currencies in which prices in attributes are expressed may be changed, for example from US Dollars $ to Pounds Sterling £, by entering an exchange rate.

Relational databases and other databases used to store attribute data usually incorporate effective editing tools. These permit a variety of operations, including searching for members of a prescribed class and then editing one by one, or assigning new values to an entire class using a single command.

9.7   FUNCTIONS FOR ANALYSING DATA

Analysing data normally comprises two principal phases:

1. Choice of data
2. Analyses of the data chosen

All GIS provide functions for analyses of data chosen and for storing the results of such analyses. Data may be selected according to:

- geographic location
- thematic content

Most GIS permit defining the criteria for selection. These are often based on SQL or are in menus with provisions for generating SQL queries. Some GIS provide predefined selection criteria. Other systems use a macro language to set up the selection criteria. Specific systems have predefined menus dedicated to the relevant applications. In most systems, selection criteria may be stored for subsequent use.

Data may be analysed at various levels:
1. Data in attribute tables are sorted for presentation in reports or use in other computer systems.
2. Operations are performed on geometric data, either in search mode or for computational purposes.
3. Arithmetic, logic operation and statistical operations are performed in attribute tables.
4. Geometry and attribute tables are jointly used to:
   – compile new sets of data, based on original and derived attributes
   – compile new sets of data based on geographical relationships

Within each of these levels, the operations used may be logical, arithmetic, geometric, statistical or a combination of two or more of these four types.

Operations may be performed on individual points or on areas, and may involve considerations of contiguity or of changes with time. Numerous operations may be performed on line networks.

The more important commands are discussed below. The functions implemented vary from one GIS to another and some GIS contain functions not discussed here.

9.7.1   LOGIC OPERATIONS

Logical searches in databases normally employ set algebra or Boolean algebra.

Set algebra uses three operators: equal to, greater than, less than, and combinations thereof:

$$=, >, <, \geq, \leq$$

These operators are included under SQL.

Practical applications include:

- Identifying extrema, such as finding attribute minima or maxima within various polygons and, as a result, delineating a new thematic layer (new

row in the attribute table).
- Selection or isolation, where particular values are selected for subsequent ranking in a new thematic layer.

Boolean algebra uses the AND, OR, NOR and NOT operators to test whether a statement is true or false. AND, OR and NOT are used in SQL. Themes A and B may give rise to the following statements:

A AND B, A OR B, A NOR B, A NOT B

These statements may be illustrated in a Venn diagram, which is a schematic representation of a set in which magnitudes illustrated by surfaces are superimposed, as shown in Fig. 9.12. The shaded areas represent true statements.

This technique is well suited to analysing geographic data. For example, potential conflicts between forestry and cattle farming can be illuminated by assigning A to potentially productive forest tracts and B to known grazing areas. The test A AND B on the two operands will identify conflict areas that can be assigned special symbols and drawn on maps.

Logic operations are particularly powerful when the relationships are complex. In GIS, logic operations may be performed simultaneously on more than two themes, and involve several operators.

9.7.2   ARITHMETIC OPERATIONS

Arithmetic operations are performed on both attribute and geometric data. All GIS support the customary arithmetic operations of addition, subtraction, multiplication, exponential, square root and the trigonometric functions:

$+, -, *, {}^{n}, /, \sqrt{}, \sin, \cos, \tan$

These operators may be used for many purposes, including assigning new thematic codes.

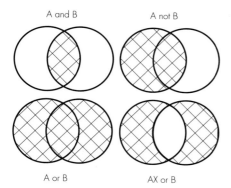

A and B   A not B

A or B   AX or B

Fig. 9.12 – Logical operations shown in a "Venn diagram". Similar operations can be carried out on the geographic data.

Typical examples include:

- Reclassification of soil types in which areas are to be converted from decares to hectares by dividing all area figures by 10.
- Distances along roads are converted to driving times by dividing all distances by a specified average vehicle speed. The result is a new set of attributes which are useful in transportation planning.

Arithmetic functions are used in all geometric computations involving coordinates, as in calculating distances, areas, volumes and directions.

9.7.3   GEOMETRIC OPERATIONS

Operations on geometric data involve the customary arithmetic operations in computations of distances, areas, volumes and directions.

In principle, distances should be measured using a uniform unit and a fixed procedure. In a raster GIS, the unit may be cell width or diagonal. In transportation analyses, results may best be stated in travel time or travel costs rather than in metres or other length units.

In many GIS, the periphery, area and centroid are automatically computed for each polygon and connected to objects as attributes for topological purposes.

With distance operations, points within a prescribed distance must be identified by comparing all distances computed with the distance prescribed.

Volumes and directions are calculated in models manipulating terrain, for instance in determining excavation volumes, terrain slopes, exposures and the like.

Flow analyses calculate geometric computations in order to find the shortest routes between designated points.

9.7.4   STATISTICAL OPERATIONS

Statistical operations are performed primarily on attribute data, but may also be effected on some types of geometric data. Most GIS support a range of statistical operations, including sum, maxima, minima, average, weighted average, frequency distribution, bidirectional comparison, standard deviation, multivariate and others.

The computation of averages requires averaging two or more attribute values and stating the result as a new attribute.

Frequency distributions are used to compile histograms – charts comprising rectangles whose areas are proportional to relative frequencies, and whose widths are proportional to class intervals. The data used to draw a histogram may also be employed to plot a curve.

Other statistical operations in common use include least-squares computations of transformation parameters from regression models, with the standard deviation as an expression of accuracy.

Bidirectional comparison involves point-by-point correlation of two themes to produce a new statistical thematic layer and hence a new attribute.

Satellite data are usually analysed statistically in dedicated image processing systems, which are often connected to GIS facilities. Some vector GIS support image processing. Multivariate operations, such as cluster analyses, are vital in image analyses. These operations assign new classes to entities on the basis of statistical selection criteria.

Pattern recognition based on statistical models is incorporated in some GIS.

9.8   MAP DATA RETRIEVAL AND SEARCH

Map data may be searched using specified criteria, using a dedicated query language or SQL. Searches start from coordinate values and, in principle, any part of a database may be searched.

Seamless databases permit searches anywhere within a geographic area. Databases based on map sheets, however, limit searches to one map sheet at a time, although contiguous map sheets may be combined for search purposes. In this case, the combined map sheets function as a partial copy of the database as a whole. For instance, a search may be conducted within an area specified by the coordinates of its lower left and upper right corners, without involving the rest of the database.

Map data usually carry thematic codes for various types of objects such as roads, administrative borders, water supply systems, buildings, properties and so on. In searching these codes, the various map themes may be "turned off and on," or shown in different colours for the different types of objects.

A typical task may require the display of all map details within the limits of x minimum = 10,000 to x maximum = 12,000 and y minimum = 5,000 to y maximum = 7,000. The query program then reads and tests all coordinate values in the database against these constraints, and retains those that lie within the stated limits.

Contiguity operations often necessitate searches in map data.

9.9   OPERATIONS ON ATTRIBUTE DATA

Inquiry in attribute data may be the primary GIS task in daily practical applications, particularly in the operational, maintenance, management and service sectors.

Inquiry calls for a man-machine dialogue in which the user employs various input/output devices and their supporting software. Tailor-made software is employed in some GIS. In others, standard routines such as SQL are employed, particularly in GIS that access relational databases.

Inquiry of and search for data may be based on logical, arithmetic operations and specific relational database functions, as with projection, selection and joining. The query criteria may be complex and cover several attribute tables.

In practice, the responses to inquiries should be displayed in a comprehensible manner, not simply as brief on-screen information. This is particularly true when attribute data are used as a reference source. Consequently, many GIS support a standard report format with fixed headings, column locations and other layout features. In addition, users may define their own report formats.

Some GIS support report formats that are tailor-made to each application, whilst others use the report function and report form storage facilities of the report generator supported by the DBMS in use.

In many cases, reports may be simple. For instance, an inquiry may be for the name and address of the owner of a property at a particular address, or for a list of all the properties registered on a specific street.

Reports in various formats may also be stored in separate files for use with other system files or for export to other systems for processing, as in statistical analyses.

Mathematical, logical and statistical operations may be performed on attribute data.

The mathematical operations include addition, subtraction, multiplication, exponential, square root and the trigonometric functions:

$$+, -, *, {}^n, /, \sqrt{}, \sin, \cos, \tan$$

The logical operations require the operators of set algebra: equal to, greater than, less than and combinations thereof:

$$=, >, <, \geq, \leq$$

as well as those of Boolean algebra:

$$AND, OR, NOR \text{ and } NOT$$

The statistical operations include:
sum, maxima, minima, average, weighted average, frequency distribution, bidirectional comparison, standard deviation, multivariate and other operations.

Attribute data may either be quantitative and expressed numerically or qualitative without numerical magnitudes. Numerical operations may be performed only on numerical data and hence mainly on quantitative attribute data. The numerical treatment of qualitative attribute data is limited to counting operations, as in determining the number of classes into which data may be ranked or how many observations fall within each class. For instance,

ATTRIBUTE TABLE

ID	Area value $
1	7.24
2	10.13
3	19.05
4	1.58
5	22.01

a)

ATTRIBUTE TABLE

ID	Area value £
1	10.86
2	15.20
3	28.58
4	2.37
5	33.02

b)

Fig. 9.13 – An arithmetic operation on attribute values. In the illustration US $ (a) are here converted to GBP (b) by multiplying by the exchange rate (1.50).

ATTRIBUTE TABLE

ID	Area value $
1	7.24
2	10.13
3	19.05
4	1.58
5	22.01

a)

ATTRIBUTE TABLE

ID	Area value $
1	7.24
4	1.58

b)

Fig. 9.14 – A logical operation on attribute values. All values below 10 are selected (b).

ID	Attributes
1	7.24
2	10.13
3	19.05
4	1.58
5	22.01

	Attributes			
ID	A	B	C	D
1	10	10.1	0.3	55
2	11		0.7	55
3		98	0.8	51
4	6	116		50

a)

	Attributes			
ID	A	B	C	D
1	10	10.1	0.3	55
4	6	116		50

b)

Fig. 9.15 – Boolean algebra on attribute values (a). Identifies all objects which contain both attribute A and B (b).

Sum $	59.91
Average $	11.98
Max. $	22.01
Min. $	1.58

Fig. 9.16 – Statistical operations on attribute values enable one to find the sum, average, maximum and minimum values in the table.

four classes, A, B, C and D may be identified and found to contain differing numbers of observations: 12 in A , 23 in B, 2 in C and 9 in D.

Attribute data are frequently processed to discern new patterns in the data.

9.10   CLASSIFICATION AND RECLASSIFICATION

Classification groups attributes according to limits set by the user. For instance, three classes may be set for the attribute of "year:" A: before 1970, B: 1971 – 1980, C: 1981 – 1990. Each object with a "year" attribute, such as a water supply pipe, is then assigned a new year-class attribute, A, B or C. Plotting the classes in distinctive symbols or colours may reveal new patterns – showing, for instance, that water supply pipes in class B (1971 – 1980) are those most subject to leakage.

Reclassification involves changing attribute values without altering geometries. Arithmetic and some statistical operations are used to assign new attribute values. In many ways, reclassification may be compared to changing colours on a map, in that the reclassified attributes may plot out in new colours.

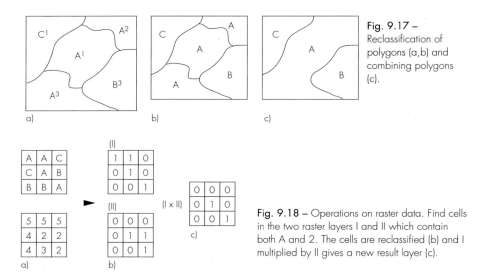

Fig. 9.17 – Reclassification of polygons (a,b) and combining polygons (c).

Fig. 9.18 – Operations on raster data. Find cells in the two raster layers I and II which contain both A and 2. The cells are reclassified (b) and I multiplied by II gives a new result layer (c).

Reclassification may be used to isolate object types. For instance, in a raster GIS, a "built-up area" theme may be isolated by assigning all other areas a value of zero. When the data are plotted out, only the built-up areas will then be shown.

The boundaries between polygons of the same type may be deleted to combine the polygons to larger units. Polygons of dissimilar types must be reclassified prior to being combined.

9.11   INTEGRATED PROCESSING OF GEOMETRY AND ATTRIBUTES

One of the simpler forms of integrated processing of geometry and attributes is to point to the location of a building displayed on screen and request retrieval of all stored information on the building. On receiving the query, the GIS searches the map database to find the building corresponding to the coordinates that have been pinpointed. Using the building i.d. number stored with the coordinates, the system then searches the attribute database for all available information, which can then be displayed or printed out.

More advanced integrated processing is also based on the condition that each object type (cultivated land, deciduous forest, protected area, etc.) is represented both in geometry and in an attribute table. The geometry concerned may be likened to a thematic map. These may be superimposed to integrate with each other and thus produce a thematic map containing information from each of the initial thematic maps. The integrated map comprises comparable units (minimum map units, MMU), and a new attribute table is compiled, as illustrated in Fig. 9.19.

Arithmetic, logical and statistical operations may be performed in the new attribute table. The geometry and the attributes may then be used to compile a new thematic map.

9.11.1   OVERLAY

Polygon Overlay

Polygon overlay is a spatial operation in which a thematic layer containing polygons is superimposed on another to form a new thematic layer with new polygons. This technique may be likened to placing map overlays on top of each other on a light table, as shown in Fig. 9.19.

The corners of each new polygon are at the intersections of the borders of the original polygons. Hence computing the coordinates of border intersections is a vital function in polygon overlay. The computations are relatively straightforward, but they must be able to cope with all conceivable geometric situations including vertical lines, parallel lines and so on. Computing the intersections of a large number of polygons may be time consuming.

If areas are stored as links in a topological model, fewer intersections need to be computed, reducing the computing time. The new intersections are identified as nodes and the lines between the nodes as links. The new nodes and links then comprise a new topological structure.

For example, consider polygon C4 in Fig. 9.20, which is a combination of polygon C and polygon 4. The system cannot associate attributes with C4 unless topology is associated with the original data or the new data. With topology for the new data, the system recognises that polygon C4 comprises line 22 and parts of line 23 and line 1. Furthermore, the system is aware that polygon 4 is on the right side of line 22 and on the left side of line 23, and that polygon C is on the left side of line 1. Hence the system now "knows" that polygon C4 is both geometrically and in terms of attributes a composite of polygons 4 and C.

Each new polygon is a new object that is represented by a row in the attribute table. Each object has a new attribute, which is represented by a column in the attribute table.

Superimposing and comparing two geometrical data sets of differing origin and accuracy often gives rise to a large number of small polygons.

For instance, two polygons representing land areas may have slightly differing geometric borders to a lake, yet on a piecemeal basis the borders may coincide. Superimposing the two polygons can then produce an unduly large number of smaller polygons.

The proliferation of smaller polygons may be automatically counteracted by laying a small zone around each line. If two zones intersect when the

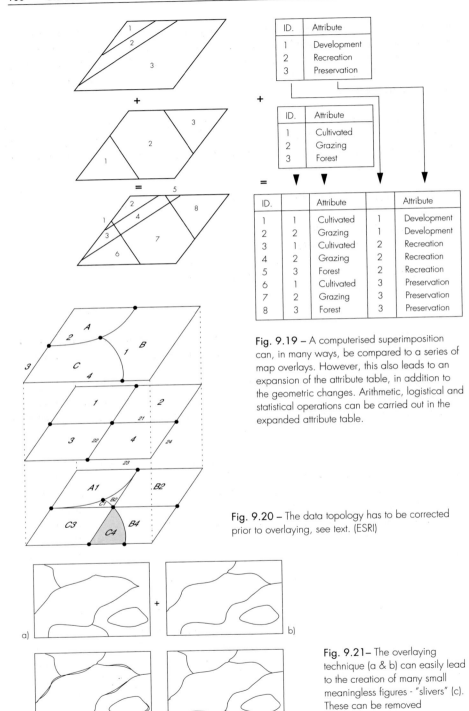

ID.	Attribute
1	Development
2	Recreation
3	Preservation

ID.	Attribute
1	Cultivated
2	Grazing
3	Forest

ID.		Attribute		Attribute
1	1	Cultivated	1	Development
2	2	Grazing	1	Development
3	1	Cultivated	2	Recreation
4	2	Grazing	2	Recreation
5	3	Forest	2	Recreation
6	1	Cultivated	3	Preservation
7	2	Grazing	3	Preservation
8	3	Forest	3	Preservation

Fig. 9.19 – A computerised superimposition can, in many ways, be compared to a series of map overlays. However, this also leads to an expansion of the attribute table, in addition to the geometric changes. Arithmetic, logistical and statistical operations can be carried out in the expanded attribute table.

Fig. 9.20 – The data topology has to be corrected prior to overlaying, see text. (ESRI)

Fig. 9.21– The overlaying technique (a & b) can easily lead to the creation of many small meaningless figures - "slivers" (c). These can be removed automatically (d).

polygons are superimposed, the lines they surround combine into one line. Smaller polygons may also be removed later, using area size, shape and other criteria. However, in practice, it is difficult to set limits that reduce the number of undesirable small polygons whilst at the same time retaining smaller polygons that are useful.

In addition to performing the overlay computations, the system can present a new image of the new structure, borders between polygons of like identity being removed to form joint polygons. This combination process may be automatic or controlled by the user.

The overall procedure for polygon overlay is to:

1. Compute intersection points.
2. Form nodes and links.
3. Establish topology and hence new objects.
4. Remove excessive small polygons where necessary and
   join like polygons.
5. Compile new attributes and additions to the attribute table.

Polygon overlay is a voluminous operation which even on the most power-ful computers may require long processing time, 15 to 60 minutes for over-laying two average map sheets.

Polygon overlay may be used to clip geographic windows in a database. For instance, a township border in one thematic layer may be used to clip all other thematic layers in order to produce a collection of data for that township only.

Overlay produces a thematic map comprising thematically comparable, homogeneous units (Minimum Map Units – MMU) and an expanded attribute table. Arithmetic, logical and statistical operations may be performed on the attributes, for instance when simulating alternatives and studying conse-quences. The advantages GIS offers in these analyses are that the number of

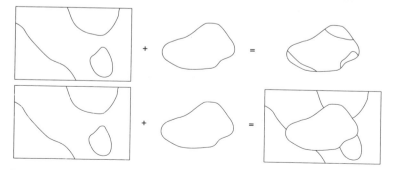

Fig. 9.22 – Different overlay operations can be used to clip out geometric windows.

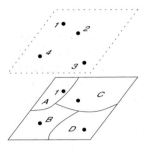

ID	Building no.	Polygon	Property
1	660	A	44/110
2	659	C	44/95
3	610	D	44/121
4	665	B	44/81

Fig. 9.23 – Superimposing points on polygons.

alternatives considered are not limited by system capacity; that all alternatives can be based on the same data; and that all available information may be used.

Points on Polygon

Just as polygons may be superimposed on other polygons, so too may points be superimposed on polygons. The points are then assigned the attributes of the polygons upon which they are superimposed.

The relevant geometric operation means points must be associated within polygons. One approach requires computing the intersection of a polygon border with parallel lines through points.
The attribute tables are updated after all points are associated with polygons.

Lines on Polygons

Lines may also be superimposed on polygons, with the result that a new set of lines contains attributes of both the original lines and the polygons.

These particular computations are similar to those used in polygon overlay: intersections are computed, nodes and links are formed, topology is established and, finally, attribute tables are updated.

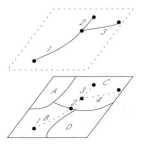

Attributes

ID	Line	Road no.	Polygon	County
1	1	Rv. 410	B	Akershus
2	1	Rv 410	C	Oslo
3	2	Rv. 9	C	Oslo
4	3	E 18	C	Oslo

Fig. 9.24 – Superimposing lines on polygons.

Raster Data Overlay

Raster data may also be overlaid. Indeed, raster overlay is often more efficient than vector overlay. The positions of the overlaid thematic layers need be tested only to see whether or not they contain cell values. The resultant cell-to-cell comparison presupposes that all cells in each thematic layer are queried, regardless of their values. The total number of cells therefore influences processing time.

The "new" composite cells are assigned attributes composed of those from the original cells. These new cells are registered as a new thematic layer.

Raster data consists of equally spaced cells of equal size (assuming that the various thematic layers cover the same area or have been modified to do so). Consequently, there is no formation of smaller erroneous polygons as with vector data overlays, and there is no need to distinguish between polygons, lines and points because all raster data comprises cells.

In raster data, attributes are usually not listed in tables as in vector data, but are represented by thematic layers. Therefore arithmetic operations and some logical and statistical operations may be performed directly during the overlay process: two thematic layers may be combined, subtracted, multiplied, and so on. If, for instance, an attribute of volume in litres is to be modified to decilitres, the thematic layer of volumes merely needs to be multiplied by 10 in each cell of an ancillary thematic layer.

The arithmetic operations on two thematic layers, A and B, produce a new thematic layer, C, through the operations:

$$C = A + B, \ C = A - B, \ C = A/B, \ C = A*B$$

Typical logical operations might be:

if $A > 100$, $C = 10$; otherwise $C = 0$
$C = $ max. (or min.) of A or B

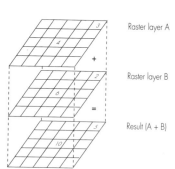

Raster layer A

Raster layer B

Result (A + B)

Fig. 9.25 – Raster data overlay is simpler than vector data overlay, and can be carried out directly on the attribute/cell values.

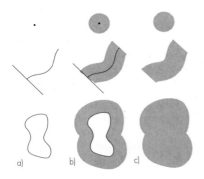

Fig. 9.26 – Buffer zones can be established around points, lines and polygons (b). The buffer zones are the basis for the creation of new polygons (c) having the attributes of the original objects, to be used in further analyses.

Some GIS support logical operations in the resultant C layer, but not directly in the original A and B layers.

As raster overlay is far more efficient than vector overlay, many GIS support functions for manipulating both raster and vector data. Vector data may be converted to raster data in order that overlaying can be performed, and the results can then be converted back to vector form.

9.11.2 BUFFER ZONES

Buffer zones are used to define spatial proximity. These comprise one or more polygons of a prescribed extent around points, lines or areas.

Many GIS support the automatic compilation of buffer zones. Here, the operator interaction usually consists of keying in specific zone parameters, such as stipulating a 50-metre zone width on either side of a road.

Buffer zones may be processed in the same way as polygons during operations that include overlay, arithmetic, logical and statistical computations in which attribute values come within the respective zones.

9.11.3   PROCEDURES IN INTEGRATED DATA ANALYSES

Integrated data analyses may be relatively complex, so fixed procedures should be employed to rationalise the various tasks, as shown below.

1. State the problem.
2. Adapt the data for geometric operations.
3. Perform the geometric operations.
4. Adapt attributes for analysis.
5. Perform attribute analysis.
6. Evaluate results.
7. Redefinitions and new analyses, if needed.

Stating the Problem

Stating problems and delineating the approaches to solutions together comprise one of the most difficult steps in using GIS analytic tools.

Consider, for instance, a typical task – the location of a recreational area which can provide a wilderness experience. The selection criteria may include:

- remoteness, at a specified distance from man-made facilities
- reasonable accessibility
- lakes and streams
- varyied topography
- a variety of vegetation

Each criterion may be analysed using layering, buffer zones and other GIS analytical functions. So before the analyses begin, the criteria to be used must be chosen carefully. Particular regard must be given to the potentials and limitations of the original data – data quality, object definitions, the bases for divisions into classes, etc.

Adapting Data for Geometric Operations

As a rule, data available from a database must almost always be modified before it can be for geometric operations. Map data, attribute data or both may, for instance, be modified by clipping a selected area out of a map database, modifying the attribute statements of the area from hectares to square kilometres, reclassification in order to reduce the number of land use classes involved.

Performing Geometric Operations

Geometric operations, which sort out the objects to be analysed, include specification of buffer zones, overlaying, search and retrieval, joining polygons, and other tasks. Each operation results in new data for further processing.

For instance, in the above example of locating a recreational area, broad buffer zones may be set up around roads and other man-made facilities. The untouched areas outside these zones may then be sorted out and overlaid with vegetation, hydrological and other data relevant to the selection.

For each task, the geometric operations must defined on the basis of the analytic criteria involved. As shown in Fig. 9.19, new polygons are formed, net topologies are established and new attributes are added.

Adapting Attributes for Analysis

Most analyses require prepared sets of geometric and attribute data, and just as geometric data must be adapted for processing, so must attribute data be prepared for the arithmetic, logical and statistical operations that need to be performed. Attribute tables must contain an adequate number of empty rows and columns for new entries. For instance, in selecting a recreational area as in the above example, a new attribute labelled "suitable characteristic" may be compiled to hold codes that indicate the degree of suitability of the various combinations of thematic layers.

Performing Attributes Analysis

Arithmetic, logical and statistical operations are performed on attribute data associated with the objects chosen in the geometric selection phase.

For instance, in the above example an analysis may be performed to identify all areas outside the buffer zones around man-made facilities. Another analysis might aim to select areas classified as moderately hilly. Yet another might try to identify minor and major streams and rivers. In all cases, data may be sorted with respect to stipulated extrema.

The final results of the operations are identifications of georeferenced characteristics that satisfy the selection criteria.

Evaluting Results

The results of analyses must always be evaluated with respect to accuracy and content. In short, do the results seem trustworthy and do they make sense? These evaluations are most easily made using maps and written reports.

Redefinitions and new Analysis

Unacceptable results must be modified, or the analyses that produced them improved and performed again. In some cases, new data may be required whilst the initial criteria may need to be changed.

Similarly, new analyses may be required in order to provide results with a greater selection of alternatives.

New analyses usually start at some suitable stage in the overall analytical process. Here again, GIS provides an efficient tool because GIS systems can process large quantities of data and perform complex computations rapidly to arrive at and compare alternatives.

Presentation of the Final Results

Analytical results are best presented in easily read maps and written reports.

9.12   EXAMPLE OF THE PRACTICAL APPLICATION OF GIS ANALYTICAL FUNCTIONS: SUPERIMPOSING AND ASSIGNING PRIORITIES TO NATURAL RESOURCE DATA

9.12.1   STATEMENT OF THE PROBLEM

In many townships in Norway, land use depends on natural resource data. However, township area planning often makes poor use of natural resource data. As a rule, extensive surveys are conducted, but further data processing, including the superimposition of data and the assigning of priorities to categories of natural resources, is often neglected.

One Norwegian township, Bygland, recently sought to improve the situation. A digital map database had been established for the purpose of area planning and natural resource themes were available on manuscript maps. Bygland sought to use GIS to process these data to natural resource data. The goal was to produce maps of selected natural themes and thereafter super-impose them to produce a map of priorities for wildlife, fishing, conservation and outdoor activities. A further goal was to produce a composite ground resource map for use in township planning and land administration.

9.12.2   PROCEDURE

Soft- and Hardware

An ArcInfo system implemented in a SUN work station was used.

Raw Data

The raw data for the task comprised existing digital data and data digitised specifically for the purpose:

- Digital basic data from 1:50,000 topographic maps: lake and watercourse contours, elevation contours and roads.
- Digital plan extract of township plan.
- Digital data for precipitation area classification.
- Manuscript maps for wildlife (reindeer, deer, moose, roe deer, black grouse, wood grouse and ptarmigan), outdoor recreation and conservation.
- Extracts of computerised registers (attribute data) from the Department of Environment:
  - Data having i.d.'s common with the manuscript maps for wildlife, outdoor recreation and conservation.
  - Fish registry data, including coordinates of fish monitoring stations.

Relating Geometry and Attributes

All natural resource themes to be related to existing computerised registers were first digitised as line information and identified by codes. For instance, the wildlife information on manuscript maps contained many overlapping symbols with consecutive numbers not associated with the species involved; see Fig. 6.3. Consequently, i.d.'s were necessary to relate the information to attribute data and facilitate subsequent sorting of the various species of wildlife.

After digitisation, the objects (points, lines and area boundaries) were linked to the existing attribute data, which also used to control the use of symbols in subsequent mapmaking.

Sorting Themes

The attribute data for wildlife, fish, conservation and outdoor recreation were sorted in layers (ARC/INFO coverage) so no areas overlapped within a layer. Typical sorting criteria were species for wildlife data and conservation topic for conservation data.

After sorting, the area data were edited in order to close all polygons for the various themes. Then the attribute data originally associated with border lines were transferred to the areas themselves. This was possible because all lines contained attributes referenced respectively to polygons on their right and left sides, and because the polygon numbers were available as numbers in the polygon topology table.

The result of the tidy up process was the basis for the thematic maps and further processing maps used to assess priorities.

Thematic Maps

Selected data from the township plan were used as background data for the thematic maps. For instance, polygons representing industrial buildings or residential areas respectively were reclassified and joined to polygons representing built-up areas. Themes intended to overlap in final map plotting were distinguished from each other.

One goal for the fish resources map was to represent fish population distributions accurately without direct reference to the discrete stock counts made in the field. This was accomplished in the following way. Firstly, the digital precipitation data for partial and complete watercourses permitted computation of fish population averages for lakes and watercourses, which provided status data independent of specie. Thereafter, a polygon layer representing the precipitation area borders was compiled from the watercourse register. Individual fish stock counts were then correlated to precipitation areas in a point-on-polygon routine. Finally, averages were computed for each area to produce the desired fish population distribution.

The first draft distributions of fish populations, and those of other themes, often revealed inconsistencies or shortcomings in the original data. The professional staffs in the relevant disciplines then supplemented or corrected the original data. With the new data, new thematic maps could be produced rapidly by replicating the processes used in producing the first drafts. This in turn required that all stages of the production process be well organised.

Use of Overlays and Buffer Zones

Differing area categories were overlayed to identify areas with overlapping themes, and union functions (NAND operations) were used in testing two layers at a time. Buffer zones were employed around such line-like information as wildlife tracks and hiking trails in order to extrapolate them to areas.

Weighting

The three most prevalent wildlife species, reindeer, roe deer and ptarmigan, were assigned priority weightings, respectively 1, 2 and 3. These weightings, which were placed in a dedicated field in the attribute table, were retained through the overlaying process to produce aggregate weightings for all polygons. Wildlife tracks, represented by buffer zones of various widths, were also weighted.

Union functions were used in successive superimpositions of layers so that the final result was a single layer containing the relevant aggregate information. Attribute fields containing the weightings for the individual thematic layers were retained in the individual polygons, so total weighting could be computed for each. High total weightings dictated that polygons should have high priority. Lower total weightings dictated lower priorities.

Unified Priorities for Wildlife, Fish, Conservation and Outdoor recreation

Weighting was also used to assign unified priorities for wildlife, fish, conservation and outdoor recreation. Each theme was divided into two classes, assigned weightings of 1 and 2 respectively. Following the overlay process,

Moos

ID	Weight
1	1
2	1
3	0
4	1
5	3

Deer

ID	Weight
1	2
2	1
3	1
4	0
5	1

Ptarmigan

ID	Weight
1	0
2	2
3	1
4	3
5	2

ID	Moos Weight	Deer Weight	Ptarmigan Weight	SUM
1	1	2	0	3
2	0	1	2	3
3	1	2	3	6
4	2	0	1	3
5	1	0	2	3
6	0	0	0	0
7	1	1	1	3

Fig. 9.27 – Use of the "Union" function to find the combined point sum for the different units in the figure, see fig. 9.28.

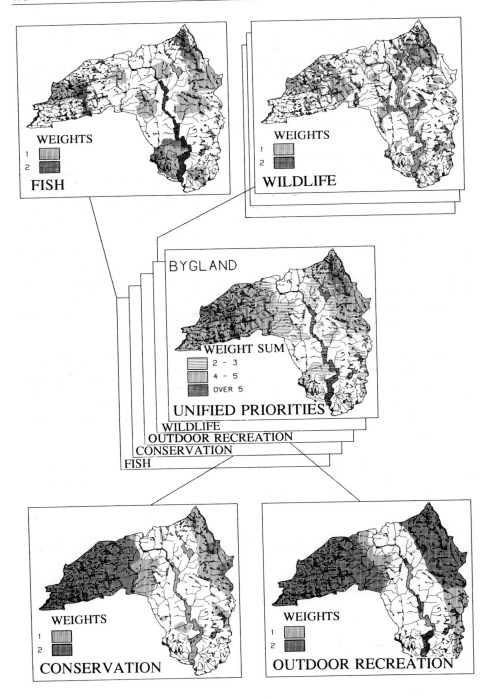

Fig. 9.28 – Unified priorities to natural resource data. (Asplan Viak)

the total weightings for the individual polygons provided the basis for plotting a priority map.

Numerous statistical computations, including sum, average and minimax, were performed in the attribute tables, both between various attributes for the same object and for like attributes for all objects in a thematic layer.

Adapting Final Results and Plotting the Priority Map

Finally, an ordinal scale of priorities of the individual areas and polygons was compiled using the weighting sums resulting from the overlay processes. Again, the final assignments identified inconsistencies and omissions. These required revision of the foregoing weightings and computations, adjustments which were made relatively easily because the weightings of the individual themes were retained throughout the various manipulations.

Once these adjustments had been completed, the final map was plotted. The ordinal priority ranks were used to control symbols assigned to the polygons, a project involving many themes that called for expert judgment in delineating borders. An important aspect of the entire "mechanics" of GIS processing was that human expertise was used continuously to evaluate intermediary and final results.

As is often the case in GIS applications, the converse of expert input in GIS processing is that GIS obliges experts to be more concise in their specifications. Another strength is that the ability of GIS to superimpose various alternatives provides rapid testing of hypotheses. GIS do not supplant professional expertise, but they do contribute to documenting it in the area planning process.

The various weightings and the assignment of ordinal priorities have much in common with the processes involved in expert systems and linear combinations, as described in section 9.15.2 and 9.15.3.

9.13   NETWORK AND RASTER CONNECTIVITY OPERATIONS

Connectivity operations utilise sets of functions that exploit the connectivity in data, and include:

- network operations
- contiguity
- spread functions
- streaming

Network operations are performed on vector data. Spread functions, streaming and some connection operations are performed mostly on raster data.

9.13.1   NETWORK OPERATIONS

Many GIS support analyses in networks, which are systems of connected lines represented in vector data. In the real world, networks consist of road systems, power grids, water supply, sewerage systems and the like, all of which transport movable resources.

Network operations are based on:

1. Continuous, connected networks.
2. Rules for displacements in a network.
3. Definitions of units of measure.
4. Accumulations of attribute values due to displacements.
5. Rules for manipulating attribute values.

GIS network operations usually include:

- displacement of resources from one place to another (route optimisation)
- allocation of resources from/to a centre

These operations are best described through the example of transport analyses.

As is the case for other GIS analytical tasks, network analyses depend on the existence of topology in the data. The system must, for instance, know which roads can accept traffic flow at intersections. Every link and every node in a network must have a unique identity and the program must contain specifications of where roads begin and end.

An analysis may include the simulation of moving resources – cars, people, refuse and so on – along the lines representing roads. The resources moved along these lines encounter resistance in the form of speed limits, road works, peak traffic, barriers, one-way streets, weight limits, traffic lights, bus stops, sharp curves and so on.

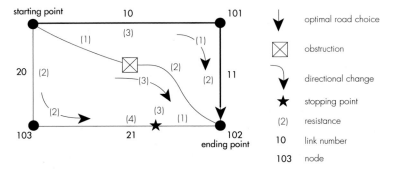

Fig. 9.29 – Network analyses can enable the system to find the fastest route between start and end points. "Resistance" (time delays) are entered as attributes and the route with least resistance is selected. (adapted from ESRI)

The network system elements comprise links, barriers, stopping points and centres. Assigning attributes to such elements permits the simulation of realistic situations. For instance, attributes representing resistance in the network may be expressed in seconds or minutes, perhaps a delay of 20 seconds at each traffic light or of one minute at each bus stop. Resistance in the network may also be computed on the basis of the accumulated distance and stipulated speed limits. A section of a road that is 1 km long and has a posted speed limit of 60 km/h represents a delay of 1 minute.

Once all the attributes have been allocated, the system can assess the movement of resources through the network. A unit of measure, such as metres, time, monetary units, etc., is then employed so that the system can evaluate alternatives and, finally, select the route of least resistance.

In this manner, transport analyses may be conducted to optimise driving time, and minimise travel distance and costs, etc. In practice, analyses usually result in several alternatives which illustrate the consequences of the foregoing simulations.

Other common network analyses include the distribution of resources from one or more centres, or the assignment of resources from an area to one or more centres.

Resources flowing through a network to or from centres must be associated with network lines or points. For instance, all pupils living in a school district are associated with their respective home addresses.

The computational processes necessary to move resources out from a centre continue to a prescribed limit. For instance, a limit of ten minutes from a centre in a street network defines those streets having a total resistance of 10 minutes from the centre. The result may be a plotted map of the streets involved.

Resources displaced from or to a centre may move along non-bilateral lines, perhaps one offering 10 seconds resistance in one direction and 15 seconds in

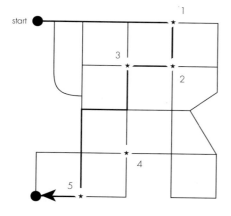

Attribute table

Stop.nr.	Passenger
1	10
2	30
3	-25
4	–
5	-15

Fig. 9.30 – The figure simulates a given bus route. At stop No. 1, ten passengers board the bus, at stop No. 2, 30 passengers get onboard, at stop No. 3, 25 people dismount, stop 4. is cut out, at stop No. 5 the remaining passengers get off and the bus carries on to the depot (end point). (adapted from ESRI)

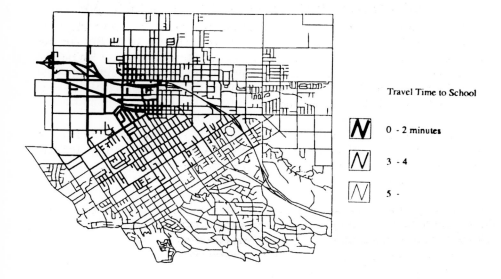

Travel Time to School

0 - 2 minutes

3 - 4

5 -

Fig. 9.31 – The displacement of resources (travel time) from a centre (ESRI).

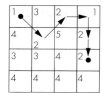

Fig. 9.32 – Optimize route location on raster data.

the other. Several centres may be involved in the same computational system.

Centres can delineate areas even further. An example might be the area within five minutes driving time of a school. Thereafter, the area delineated may be employed either in a proximity analysis of the objects it encompasses, or in an overlaying function.

9.13.2 CONNECTIVITY OPERATIONS ON RASTER DATA

A few GIS support connectivity operations in raster data, a process which requires discrete cell-by-cell displacements, originating from a starting point. The process paradigm involves the choice of units of measure and of other measurement parameters. The attribute values of the cells are accumulated in transit of each cell.

Connective operations may be used to determine travel distances in a road network, to identify areas of given shapes and sizes, and to compute precipitation run-off from a terrain surface.

9.14 CONTIGUITY OPERATIONS AND INTERPOLATION

All GIS support some form of contiguity operation, in which attribute values are assigned to new points on the basis of the values of the existing neighbouring points or observations. In ordinary circumstances, contiguity manipulations involve routines for area search and interpolation. Contiguity operations are considered to include topographic functions associated with manipulations in digital terrain models.

Search operations usually take place within an area, usually from a position to a new point, as within a circle around a point. Existing attribute values within the area are treated in accordance with the criteria for determining the values of new points.

The procedure for contiguity operations usually comprises four main steps:

1. Identify a base point.
2. Define/compute the search area.
3. Select/search for objects.
4. Manipulate the attribute data in accordance with selection criteria.

For instance, the manipulations of attribute data may comprise statistical computations, including:

- sum of all values
- average of values
- greatest or least value
- interpolating values from those of neighbouring objects
- identification of the most frequently occurring value
- statistical distribution of values

Area search and the subsequent computation of elevations of grid points from elevations of points within grid cells are used to compile digital terrain grid models.

Interpolation may be based on selection criteria other than area search or may be processed in mathematical functions, for example by linear interpol-

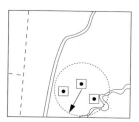

Fig. 9.33 – An area search operation can be carried out from a given point and within a circle of a given radius.

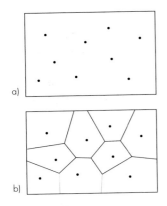

a)

b)

Fig. 9.34 – The Thiessen polygon. The polygon is created around a point with the result that the polygon's border will be located equidistant from the adjacent point.

Fig. 9.35 – A contiguity operation using step-by-step window scanning of raster data. The value of the cells within the window can be manipulated in different ways.

3 x 3

1	2	2	2	2	2	2	2
1	1	2	2	2	2	2	2
1	1	3	3	3	2	2	3
1	1	3	3	3	2	3	3
4	4	4	4	4	3	3	1
4	4	4	4	3	3	1	1
4	4	4	3	3	1	1	1
4	4	3	3	1	1	1	1

ation between two or more known points. Borders may also be interpolated around points to form surfaces. These surfaces may be formed using Thiessen polygons, in which the borders around a point are delineated equally distant from all neighbouring points.

The search criteria for attribute data may be stipulated simultaneously with the delineation of the search window, which can simply be a specification of the coordinates of the upper left and lower right corners. For instance, in a theme covering properties, the search may be set up for all residences on lots larger than five decares. Such searches often combine data from several thematic layers, thus necessitating searches amongst them.

Another area search criterion might be a random polygon, perhaps all the cultivated fields within a given window that correspond directly withing a given polygon. Hence, in many respects, contiguity operations resemble overlaying.

Contiguity Operations on Raster Data

Raster data are well suited for contiguity operations. A raster GIS can assign new cell values independent of the values of the neighbouring cells of the original layer. This may be accomplished, for example, by step-by-step scanning using a three-by-three cell window. The value of the cell in the middle of the window is computed from the average of the values of all nine cells viewed , or by other calculations.

Contiguity operations may also be used in networks, for instgance by determining areas to be searched on the basis of stipulated driving times from specified points in a road network.

9.15   GIS ANALYTIC MODELS

Several theoretical models have been evolved to exploit GIS analytic functions. The three prime examples are cartographic algebra, expert systems and linear combinations.

The models must be regarded as aids to expert judgement because the drawback in all of them is that they are based on theoretical assumptions which cannot cover all practical situations.

9.15.1   CARTOGRAPHIC ALGEBRA

Cartographic algebra is based on the assumption that a set of simple operations can be found and sequentially joined to comprise relatively complex modelling. This process starts with an existing set of attribute tables, or raster data which are then processed in a sequence of operations, each of which produces a new column in a table. Finally, the attribute sought is compiled. It may be represented by a flow chart in which the thematic map is successively processed until the final thematic map is produced (G.H. Strand).

Cartographic algebra may require the following operations:

- reclassification
- averaging
- maximising
- subtraction
- addition
- scattering
- streaming

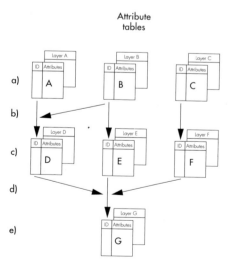

Fig. 9.36 – Cartographic algebra. Thematic layers A, B and C with their own attribute tables (a) are manipulated by simple operations (b) to a new thematic layer and attribute tables (c) which are, again, by simple operations manipulated to (d) one result layer and a result attribute table (e).

The success of the model is contingent upon the manipulated data being of a form suitable to the operations concerned. For instance, attribute values arranged in classes cannot be added or subtracted, and are therefore unsuitable.

9.15.2   EXPERT SYSTEMS

GIS are increasingly implemented with ever more "intelligence." One "intelligent" approach is that of the expert system, in which analytical results are automatically assessed in terms of criteria entered as ancillary information, or attributes. On the basis of the criteria entered, the system assigns priorities to various combinations of attributes and then provides an output comprising recommended choices and measures.

In GIS, each attribute type may consist of a number of attribute values. For instance, "soil quality" may have three quality classes, "conservation" may have two conversation classes, and "quaternary geology" may comprise three types of geologic deposits. The attribute values of these three attribute types may be combined to 3 X 2 X 3 = 18 possible combinations. One or more experts may then rank the 18 combinations with regard to a particular goal, such as development. All combinations including conversation classes are ill suited to development. Some combinations may be unworkable for technical or other reasons, and hence must be discarded.

The evaluations of the experts are then arranged in a table in which the 18 combinations are ranked qualitatively, such as into classes of "very suitable," "suitable," "moderately suitable," "moderately unsuitable," and "very unsuit-

Fig. 9.37 – Expert systems. The experts' criteria in table (a) are combined with attribute table (b) for an automatic evaluation of the result of an overlay operation, see fig. 9.19.

a)

Attribute	Attribute	Priority
Cultivated	Development	9
Cultivated	Recreation	6
Cultivated	Preservation	7
Grazing	Development	3
Grazing	Recreation	1
Grazing	Preservation	8
Forest	Development	4
Forest	Recreation	2

b)

ID.		Attribute		Attribute	Priority
1	1	Cultivated	1	Development	9
2	2	Grazing	1	Development	3
3	1	Cultivated	2	Recreation	6
4	2	Grazing	2	Recreation	1
5	3	Forest	2	Recreation	2
6	1	Cultivated	3	Preservation	7
7	2	Grazing	3	Preservation	8
8	3	Forest	3	Preservation	5

able." The table can then be entered in the GIS and serve as a reference for every record (MMU) in the attribute table that results from GIS analyses. The expert recommendations can thus be entered as a new attribute, or column in the attribute table, and be queried whenever reports are exported.

The process clearly implies that the answers provided by an expert system can be no more reliable than the professional expertise available whilst compiling the evaluations entered.

9.15.3   LINEAR COMBINATION

The linear combination model is used in impact assessment and in suitability analysis.

The starting point is a scale, as from 0 to 5, which is used to rank features with respect to a stated application. The attribute values of all attribute types are first coded by a knowledgeable expert. Rank on the scale then indicates the mutual relationships of the attribute values. In the above expert systems example, for instance, the three categories of soil quality may be assigned 1 for low quality, 3 for medium quality and 5 for high quality. Corresponding rankings may be assigned to the conservation and quaternary geology attributes. The result is that all attributes are ranked.

In practice, the various types of attributes will have varying impacts on the end product of the analysis. Conservation, for instance, may weigh more heavily than geology in selecting an area for development. Consequently, the various types of attributes must be weighted to indicate their relative importance.

All records (MMU) in the attribute table are then weighted according to the attribute type involved. The result is a set of weighted features, which may then become new attribute values in the attribute table.

9.16   DIGITAL TERRAIN MODELS

Digital terrain models may be used in various operations which are partly included in the operations discussed above, as contiguity.

Elevations

Using elevation data stored as a point cloud, the elevation of a random point can be computed as the weighted median of the surrounding points, with the closest points having the greatest weight. Neighbouring points are searched within a specified area.

Conflict and other mathematical functions more complex than the weighted median may also be used (Kriging).

Fig. 9.38 – Area search.

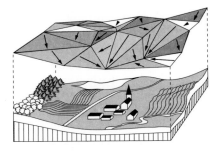

Fig. 9.39 – Slope calculation in a TIN model.

The elevation of a new point in a grid model may be computed as the elevation of a plane passed through the closest grid points. Elevations may also be computed from the faces and corner elevations in a Triangulated Irregular Network model.

Slope and Slope Direction

A terrain's slope and direction may be computed relative to a plane through the elevation model points.

Slope degrees and direction are useful in simulations of precipitation run-off and in defining drainage areas.

The triangle terrain model, which comprises sloping triangular surfaces, is ideal for simulating terrain surface conditions. Slope and slope direction are computed directly from the coordinates of the corners of a triangle.

Slope is usually expressed in degrees or percentage terms with respect to the horizontal. Slope direction is almost always expressed in degrees from North.

Contours

The contour lines of constant elevation may be interpolated directly from the digital elevation model. The program used identifies the x and y coordinates for new points having the same interpolated elevation, and then connects them together to form contour lines.

Many methods are used. In the Triangulated Irregular Network (TIN) model, triangle faces intersecting a stipulated contour elevation are identified, and the points of intersection determined by linear interpolation. In elevation data stored in the form of point clouds, triangles are first formed, then treated as are those of the triangular terrain model.

In grid models, the grid points on either side of a contour line are identified. A plane through the points or linear computations then locates the new point on the contour line.

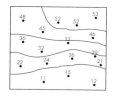

Fig. 9.40 – Calculation of isolines (contour lines) in a point cloud of spot heights.

In all cases, the contour interval can be chosen to suit the accuracy required. The various approaches to generating contour lines can produce differing results on plotted maps.

Volumes

Accurate surveys of excavation volumes are crucial in planning road works, major industrial buildings, and so on. Soil borings are used to calculate cover quantities, such as down to bedrock, and replacement quantities, as down to usable soil. From these quantities, haulage can be calculated down to the project level.

Quantity surveys may be conducted in GIS by entering data in a terrain model program, which includes the extent of the excavations in elevation and ground plan, soil conditions, slopes for planned levelling, volume expansions due to blasting, etc. The quantity survey computations are based on calculations of cross-sectional areas multiplied by the intervening distances between cross-sections.

Visualisation Techniques

The results of excavations may be visualised by excising a rectangular area in the terrain model and viewing it at an angle from above. Users are usually free to select the viewing angle that best displays the features sought. Profiles and grids in the area are plotted to suitable densities. Regular grids are best for the purpose, so several Triangulated Irregular Network (TIN) models support regular grids for plotting.

Drawing straight lines between successive grid points often produces a rough, serrated terrain surface. Consequently, some systems support splines, or curved line segments, for plotting.

Perspective drawings employing profiles are usually based on isometric perspective images of terrain. That is, the units of measure are fixed along the axes, and the foreground of the image has the same scale as the background. Central perspective produces a more realistic image of terrain, but does not permit measurements because scales vary in plots.

For the best replication of terrain, the program must delete profile lines and surfaces hidden by terrain rises as seen from a side view. The deletion

Fig. 9.41 – Perspective drawing.("Terrapartner" ©, Viak System)

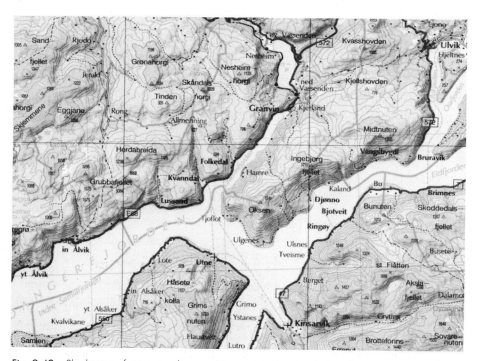

Fig. 9.42 – Shadowing of a topographic map (Norwegian Mapping Authority)

process calls for the sequential sorting of lines, starting from the observation point and working backwards. The lines are tested one by one for visibility. Computationally, the process is cumbersome and time consuming.

Shading and Draping

A terrain model may be used to produce automatically relief maps which use shading to effect the appearance of the third dimension of height. These processes demand the computation of shaded areas from an assumed solar position.

The simplest approach is to colour the cells of a raster model. The colours can be varied with elevation, as on ordinary topographic maps. The colour intensity can be varied with cell slope, to give the impression of sun and shadow.

Draping a digital aerial photo, satellite image or thematic map over the surface of a terrain model produces a more realistic terrain image. However, draping two-dimensional images over three-dimensional surfaces poses problems because the two-dimensional image does not contain enough pixels to cover the terrain surfaces. The system must then compute pixel values for surfaces not covered, which deforms the two-dimensional image.

Accuracy

High accuracy is required in all terrain models to be used for engineering purposes. The accuracies of terrain descriptions are determined primarily by the accuracies, amounts and distributions of basic data; by the method of generating the model grid and triangular surfaces; and by the method used to interpolate between points in the model.

For a grid model, the following accuracies are typical:

Source	Accuracy in elevation and ground plan
• surveying	± 5 cm.
• photogrammetric data from 1:6,000 images	± 20 to 30 cm
• digitised 1:1,000 maps	± 50 cm

In those models in which cells and profiles are recreated from a point cloud with each computation, accuracy depends on the cell or profile density. Profiles must be closer to each other in order to represent more rapid terrain variations, but greater profile densities naturally call for the processing of greater amounts of data.

Miscellaneous Applications

The principles of interpolation and visualisation in digital terrain models may be applied to other types of three-dimensional data. For instance, the third dimension in georeferenced data may be time, population data, farm productivity, geophysical measurements and so on, see Fig. 6.20. In visualisations, elevations may be replaced by any numerical values related to the observation points. Likewise, interpolation between contours may be performed or perspective drawings plotted.

9.17   FUNCTIONS FOR "ENGINEERING GIS"

Geographic data are often used in solving such engineering problems as locating basic points, surveying properties and planning various phases of roads, water supplies, sewerage, residential areas and so on. These tasks have been accomplished previously using stand-alone PC programs, but the current trend is to integrate them into Urban GIS.

Surveying and other Measurements

Surveying programs usually include functions for computing coordinate geometries based on in-field measurements, with interfaces to various surveying instruments. For engineering applications, they may be extended to compute the intersections between lines, and between lines and circles and so on. Then, they must support computations of various geometric figures so that exact geometries may be computed.

Other operations may also be performed. The lay out data of plans may be computed, various map projections may be manipulated and transformations made between projections. Often, the systems can produce standardised deed or other maps in local or standard formats.

Some systems support the method of least squares, as well as automatic quality designations in the form of error ellipses. Other systems can combine GPS vectors with conventional measurements.

Manipulating Terrain

To aid engineering work, digital terrain model systems have been developed to manipulate simultaneiously as many as ten separate terrain layers, such as loose fill quantity, bedrock and various geologic strata. Volumes for cut and fill operations are automatically computed and documented in maps, perspective drawings, profiles and cut specifications.

For instance, when a new road has been aligned, it may be line designed with exact curve radii, transition curves, lengths and cross-sections. There-

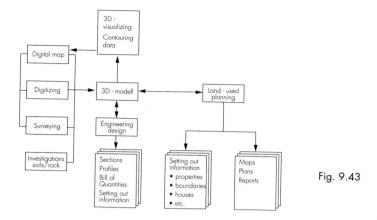

Fig. 9.43

after, its plans may be plotted with centre lines, driving lanes, shoulders, cuts, fills, building lines, etc. These data are then used to set out lines in terrain. The documentation concerned may be standardised, for instance to the standards of a road works commission.

As visualisation prior to construction is desirable in many engineering applications, some GIS terrain models are integrated with three-dimensional Computer-Aided Design (CAD) systems. Such a combination can present realistic images of the planned construction, perhaps showing a complete building, with facades, windows and interiors surrounded by graphic symbols for trees, cars and other extraneous objects.

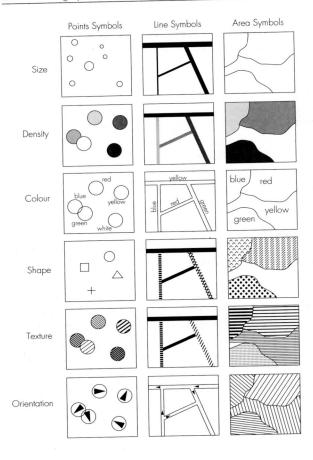

Points Symbols   Line Symbols   Area Symbols

Size

Density

Colour

Shape

Texture

Orientation

Fig. 10.1 – In principle there are six different graphic parameters which can be used for symbolising geographic phenomena - shape, orientation, colour, texture, density and size.

These six parameters differ in their application – for example, illustrating the qualitative characteristics of various object types, as with tree species – pine, fir or beech, etc. – or the quantitative characteristics of various attribute values, as with degree of sulphurous emissions, number of inhabitants, age, etc.

Size

Varying the size of symbols is the most convenient way of illustrating variations in the quantity represented by a symbol. The symbol's size can then be made a direct function of the magnitude measured. Diagrams, including the frequently-used "cake diagrams," seldom convey information successfully on maps.

However, differences in size are not always so readily perceived by human users. As with most areas of human perception, the difference in magnitude required to elicit perception of it is proportional to the initial magnitude involved. Thus, the diameter of a circular symbol representing a given amount must be nearly tripled in order to convey the impression of doubling it.

Density

Grey scale, or variation of density, is used mainly to illustrate quantitative data which are ranked. As is the case for all cartographic parameters, there is a lower limit to the size of the differences which can be perceived by a user. Grey scales, for instance, should be divided into no more than ten sections, lest neighbouring sections be mistaken for each other. However, if dot rasters are also employed, the number of classes represented may be increased to upwards of 20.

To be perceived as regular, the steps on a grey scale must be irregular. This is because regular increases of blackness, for example from 10% to 20% to 30%, and so on, are perceived as being irregular. A ten-step scale perceived as comprising equal steps should have a progression of white (0%), 9%, 19%, 31%, 45%, 60%, 74%, 84%, 91%, 100% black (Baudouin and Anker 1984).

The properties of density are illustrated in the map of Fig. 10.2. The immediate impression is that various magnitudes are presented, but the data shown on the map are actually different types of forest growth, which are qualitative. The darker areas are more easily perceived than the lighter. This shows that variation in density results in differing visual impressions which clearly separate the areas represented. If, however, the forest growth areas in Fig. 10.2 had been graphically distinguished from each other through differences in symbol orientation, they would not have seemed to represent data ranked by magnitude.

Maps in which symbols cover areas are often termed choropleth maps, from the Greek choros (place) and plethos (magnitude).

Colour

Colours are effective and pleasing in appearance and are particularly well suited to distinguishing various qualitative phenomena from each other. Unfortunately, colours are also the most frequently misused of all the carto-

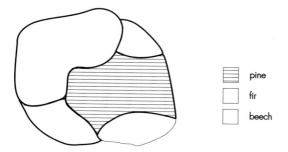

pine

fir

beech

Fig. 10.2 – An example of the incorrect use of density as a graphic parameter. The impression might easily be created that magnitudes are being presented on the map, whereas, in fact, the data shown are qualitative rather than quantitative. (Bjørke).

graphic parameters. Although colours are often coded to scales and used to indicate quantitative properties of areas, such as population densities, monthly temperature means, degrees of industrialisation and the like, colours themselves carry no subjective implication of rank or order. Red, blue, green and orange might well represent the values 10, 20, 30 and 40 of some measurable quantity shown on a map, but users can only interpret the rank of a coloured area correctly if the map has a colour code scale that indexes the colours used to the quantity involved. Densities, however, directly imply variations of magnitude: darker means greater. Therefore, the tone of a single colour, which can be varied by adding white or black to the colour, usually conveys relative magnitude better than the variation of base colour.

Variations in the tone of a single colour are, in fact, often more obvious to users than are differences in colour. In principle, colours could be varied continuously because the spectrum of visible light is itself continuous. However, it is virtually impossible to achieve a continuous colour change on printed maps.

The use of colours also permits the exploitation of common associations. For instance, red, which is commonly associated with danger, might be used to indicate dangerous areas on a map.

Shape

Varying geometric shapes are best suited to indicating qualitative differences. They convey no overall impression, but may be used to convey details. Shapes are usually noticed last of all the cartographic parameters.

Texture

Very occasionally, variations in texture are used to show quantitative differences, usually by varying the printed dot density of a fixed percentage mix of two colours, such as 70% black and 30% white.

Orientation

Line symbols in various orientations are best suited for illustrating qualitative differences.

10.2   SELECTING MAP SYMBOLS

In Fig. 10.3, the cartographic parameters are assessed in terms of their perceived visibility, their ability to distinguish between various phenomena, and to connote rank or order.

In maps containing several cartographic parameters, patterns distingui-

Visual variables	Phenomena ability
size	quantiative
density	order
texture	quantiative
colour	quality
orientation	quality
shape	quality

Fig. 10.3 – Illustration of the relationship between visual graphic parameters and their abilities.

shed by symbol size and grey scale seem more visible than those distingui-shed by symbol shape and size. So the cartographic parameters may be ranked according to impact.

The cartographic parameters may be combined to display several pheno-mena simultaneously. For instance, colours may be varied among varying dot or line patterns. Patterns that combine the cartographic parameters are fre-quently used to present data that are arranged in a particular order.

The graphical display of information can be successful only if the display parameters have the same mutual order, similarity and magnitudes as the variables illustrated (A. Bauduin 1984).

The choice of map symbols is seldom straightforward. The following guidelines (C. Palm 1988) are useful:

1. Select point, line or area symbols.
2. Consider the final application of the map (tourism, documentation, planning, etc.) and the capabilities of its users.
3. Consider the characteristics of the cartographic parameters.
4. Use symbols of equal impact to represent equally important variables.
5. Use related symbols for related phenomena.
6. Consider visual phenomena and their characteristics. For instance:
   • Symmetrical symbols (roads, railways, etc.) often dominate.
   • Symbols appear smaller when they are surrounded by larger symbols, and vice versa.
   • Whenever two areas adjoin without a border line, a light area appears lighter and a dark area appears darker.
7. Heed tradition (red means danger, the sea is blue, etc.) and existing standards.

The most common errors in cartographic presentations include:

• Incorrect use of cartographic parameters.
• Background colour too strong in relation to the most important information presented.
• Excessive number of dissimilar themes presented.
• Inadequate legend or instructive material.

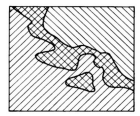 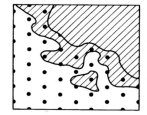

Fig. 10.4 – In this representation of the diversification of language areas and overlapping of language groups a) gives the unfortunate impression that the overlap area represents a new language group, whereas b) gives the correct visual impression. (Anker and Bouduin).

Maps on which several phenomena are presented often contain areas where adjacent phenomena overlap. Graphic parameters must then be chosen carefully, so that users readily comprehend the overlaps and do not interpret them as separate phenomena; see Fig. 10.4.

10.3   THE POTENTIALS AND LIMITATIONS OF GIS IN CARTOGRAPHIC COMMUNICATIONS

The ability of a GIS facility to employ cartographic parameters is determined primarily by the constraints imposed by the software systems of its screens and plotters. These programs usually contain pre-defined symbol libraries.

Users may often define their own symbols by digitising symbols drawn or printed on paper or by using software tools to construct their own symbols.

As illustrated in Fig. 10.5, large selections of different types of lines, line thicknesses, special symbols and fonts are available.

Some systems support printouts of defined attribute values, such as measured values and names of measurement stations, or permit attribute values to direct the plotting of selected symbols. The size or colour of symbols may be related to the attribute value by checking with a reference table (Look-up table).

Density or solid colours may be used to indicate phenomena covering areas on a map. Colours may, for instance, be assigned numerical codes which are entered into the plotting codes for the map polygons, which are in turn activated when the map is plotted.

Most GIS facilities support a range of fonts, either in basic system software or in plotter programs.

The screens and plotters of a vector GIS facility cannot plot solid colours in areas, but employ line patterns. Their support of colours and grey scales may also be limited. Raster displays support colour assignments to each and every dot, so raster plots can include solid colours. However, raster resolution may

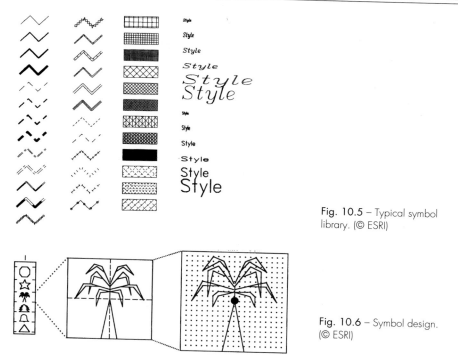

Fig. 10.5 – Typical symbol library. (© ESRI)

Fig. 10.6 – Symbol design. (© ESRI)

limit the detail of symbols that can be drawn. Similarly, variations in texture may be difficult to achieve.

In general, screen resolution sets the upper limit to on-screen enlargement of a map: the greater the resolution, the more a displayed map can be enlarged.

Line plotters cannot support solid colours, but can produce fine lines for direction symbols and hachure. Their colour capabilities are set by the number of pens available. The drawback of pens is that the quality of the lines they produce may vary because the ink in pens dries, and nibs wear. Moreover, the line's width and density may vary if the flow of ink cannot keep up with the speed of the pen at higher drawing speeds.

Plotters, whether colour raster plotters or film printers, seldom replicate on-screen colours exactly. This limitation may be vexing, particulary when colour nuances are vital, as in the processing of remote sensing images. In such cases, extensive calibration may be required.

Raster plotters are best suited for producing graphic effects because their characteristics are similar to those of raster screens. The quality of the printed product depends on plotter resolution.

Both raster screens and raster plotters may be limited in their ability to fill an area with solid colour, for example by not accepting polygons larger than those that can be delineated by borders comprising a maximum of 2,500 points.

Some systems support overall changes, for instance by altering the size of

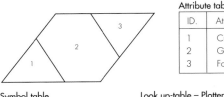

Attribute table

ID.	Attribute
1	Cultivated
2	Grazing
3	Forest

Fig. 10.7 – Look-up table.

Symbol table

Symbol no.	Symbol
Cultivated	2
Grazing	1
Forest	4

Look up-table – Plotter

Symbol no.	Symbol
1	
2	
3	
4	

Look up-table – Screen

Symbol no.	Symbol
1	Read
2	Green
3	Yellow
4	Blue

symbols and the width of lines whenever the scale of a map is changed.

Beautiful final products may, of course, always hide poor quality data. Data uncertainty is not easily shown graphically. Uncertain boundaries between polygons can be shown by dotted lines, double lines, a line darker in its middle and lighter towards its edges, or as a zone of a particular colour or no colour at all.

Maps are static presentations that are not particularly suited to showing changes with time. But animation techniques can bring on-screen maps to life. For instance, maps may be displayed rapidly in sequence in order to illustrate a particular trend, such as the day-by-day spread of radioactive fallout from Chernobyl during the first two weeks following the disaster.

In short, a GIS exploits its analytical prowess to rapidly generate a wide variety of maps from the same data. However, GIS maps have cartographic limitations because few aesthetic capabilities are supported in GIS. Furthermore, current GIS do not have the wide range of fonts and symbols available to conventional cartographers, whose uses of fonts and symbols are limited only by individual imagination. GIS supports an enormous range of colour hue, tone and intensity, but in practice the ranges available may be more restricted in producing hard-copy maps. The GIS user cannot manipulate the overall aesthetic appeal of a map as can the conventional cartographer.

One of the underlying reasons for this disparity is that the art of map-making (and map printing) has developed over the centuries. The art, like the artisan, is mature. GIS is still young. Consequently, its cartographic aesthetics are relatively crude, and the disciplines involved are less mature than those of conventional cartography.

So there are disparities. The skills of a GIS user are related to the analytical manipulation of spatial relationships, whereas the skills of a conventional cartographer are related to the representation of spatial relationships. The GIS user relies on the technological end products of the sciences, as computational methods, mathematical operations and the like. The conventional cartographer relies more on the arts, which involve perception and are therefore integral in aesthetics and communications.

11.   INITIATING GIS

11.1   INTRODUCTION
11.1.1   BACKGROUND OF THE CENTURIES

In many ways, the evolution of the applications of geographic information mirrors history itself. From the earliest charts to the most recent atlases of the universe, maps have always reflected not only a contemporary knowledge of the world but the contemporary technologies employed to gather it. From the earliest Roman census compiled for taxation purposes to the Domesday Book to modern demographic studies, overviews of populations and their distributions have recorded stages in the development of commerce, trade and government. Conversely, many of the quests for improved georeferenced information have spawned changes that have affected history. The search for a reliable method of measuring longitude during the 16th, 17th and early 18th centuries triggered the first government sponsorship of scientific research projects. Similarly, research into ballistics during the 1930s and 1940s led to the development of the digital computer, one of the earlier examples of civilian benefits resulting from military R&D spending.

Initially, individual mapmakers expended considerable time for limited gain in the knowledge of their immediate areas, although maps were established as useful sources of geographic information. Through the centuries innovation spurred progress. Accurate longitudinal measurement, the theodolite, precision printing, photogrammetry, aerial photography, satellite remote sensing and a host of lesser technological breakthroughs all prompted progress. As a result, the curve of qualitative cartographic improvement ver-

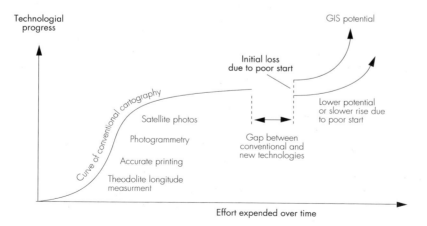

Fig. 11.1 - Qualitative development in cartographic disciplines.

sus the effort expended to gain it soared - although the curve subsequently tailed off again as the effort given to development produced fewer overall gains.

Map sheets produced by traditional manual means in the late 20th century may be more attractive and more accurate than those of the late 19th century, but updating them still requires a considerable investment in man-hours. The details of a public utility network of the 1990s might be more involved and far more extensive than those of the same network of the 1890s, but the filing of information, on sheets of paper in folders in file drawers, may have remained unchanged. So the human element - the speed and accuracy with which a cartographer can draw, the time required for a clerk to find and fetch a file - is clearly the ultimate constraint.

Computerised Geographic Information Systems, however, can dramatically change this overall picture because it dispenses with much of the human intervention required for the conventional acquisition, manipulation and storage of geographic information.

An organisation dealing with georeferenced information may initiate GIS to realise a gain, but as implied in Fig. 11.1, effort (in time and money) is required to switch from conventional means to GIS. Initially GIS may offer little or no qualitative gain; in some cases, if incorrectly or inexpertly applied, it may actually degrade product quality. Bridging the effort gap successfully is therefore the first step towards realising the benefits afforded by GIS.

11.1.2   BRIDGING THE GAP

Bridging the gap invariably means switching to a new technology and changing procedures. In modern business language, the first steps to initiating GIS in an organisation call for a number of strategic choices. For many high-tech approaches, the benefits accrued are related to these initial choices. The Nordic KVANTIF retrospective survey concluded that very few organisations regretted the switch to GIS, but that inretrospect many felt their GIS activities could have been started more profitably in other ways.

A GIS facility cannot be bought "off the shelf." It is rather an assemblage of various items of equipment that can become a useful tool only when it is properly placed in an organisation and supported by expertise, structured data and organisational routines.

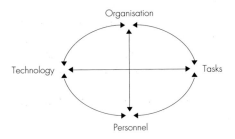

Fig. 11.2 - Successful initiation of GIS depends on a prudent balance of technology, organisation, personnel and tasks.

10.   CARTOGRAPHIC COMMUNICATION

10.1   THEORETICAL FOUNDATION

As discussed in the preceding chapters, Geographic Information Systems are not just automatic map makers, but are capable of a wide range of tasks. Nonetheless, maps, whether displayed on screen or printed on paper, are a primary GIS product for end users.

Regardless of how a map is produced, it is a visual communications medium and, as with other visual communications media, appearance affects how a map is perceived and, consequently, how readily the information it contains is interpreted by the user. This is one of the reasons why traditional cartography is a venerable craft in which artisans train for long periods to master the aesthetic appeal that is necessary to ensure that maps communicate information to users. Even the most skilled cartographers, though, are limited by the drawing implements at their disposal. Computerised GIS are not so limited and are therefore capable of producing a far greater variety of maps than can be drawn by cartographers, even when using the best hand tools of their craft.

This capability has the enormous potential of enabling less skilled persons to produce truly professional maps, but it also has the drawback of permitting the production of artless maps that are at best unattractive and at worst misleading. Thus far, GIS facilities have no cartographic intelligence that can guide operators in the choice of map symbols and other graphic effects. Consequently, those who operate GIS facilities should be reasonably knowledgeable in all aspects of cartographic communication.

From a communications viewpoint, maps may be divided into two main groups: the viewable and the readable. Viewable maps are devised to convey messages to users immediately. A typical example is the thematic map. Absorbing information in readable maps, however, requires more time because, generally, they must be interpreted, usually by study and reference to their legends. Ordinary topographic maps in various scales, such as 1:1,000, 1:5,000 and 1:50,000, are examples of readable maps. So, whilst viewable maps usually are the most effective communications media, readable maps are the most effective storage media for geographic information, although their graphic appearance is also vital (J.T. Bjørke 1988).

10.1.1 CARTOGRAPHIC PARAMETERS

In principle, cartographers command six graphic parameters in symbolising geographic phenomena: size, density, texture, colour, orientation and shape; see Fig. 10.1.

GIS – CHAIN

Expertise | Structured data | Organisation | Hard– / Software

Fig. 11.3 - The interlinked chain of a GIS facility.

The tasks and issues involved in initiating GIS may be classified as either organisational or technological. As for other major facilities, the process leading up to the implementation of a GIS usually comprises several steps:

- Studies
- Pilot project
- Main project plan/system implementation

The budget of the preparatory phases should, of course, reflect the cost and complexity of the envisioned system. An investment in the preparatory phases of between five and 10 per cent of an overall project budget is usually adequate to minimise project risks.

The scope of a GIS implementation project is related to the final impact of GIS on an organisation. One may distinguish two categories according to impact:

1. Minimum impact. Within the organisation, GIS is seen as a useful new tool to be acquired, much as a new computer or telephone exchange. Typically, the GIS users may be within a division of a larger company or a bureau of a governmental agency.

2. Major impact. Initiating GIS changes the way in which the organisation operates. Consequently, the problems addressed affect virtually all phases of the organisational hierarchy and operational paradigms. Typical organisations include municipal authorities, public utilities and cartographic agencies.

The demarcation between these two categories is a blurred zone rather than a sharp line. GIS activated in a small bureau of a large governmental agency may radically change that bureau. For the bureau, the impact is major: for the parent agency, GIS is merely a tool that enables one of its many bureaus to accomplish its task. Conversely, a large agency in a small country might have to be reorganised completely in its conversion to GIS. In terms of equivalent efforts in larger countries, however, its efforts may be small. Hence the qualifier "from the users' viewpoint" might be applied to all assessments of the impact of initiating GIS. The bulk of the following discussion is therefore devoted to the installation of GIS in organisations for which it represents a major impact.

A third category of organisations spawned by GIS, such as national geographic databases, are discussed in Section 11.5 of this chapter.

11.2   THE ORGANISATIONAL EVOLUTION OF A MAJOR GIS IMPLEMENTATION

As pointed out in preceding sections of this chapter,GIS differs considerably from its conventional predecessors. Consequently, the effect that exploitation of GIS has on an organisation often requires restructuring. The prudent, realistic approach usually requires a balance between remodelling the organisation to accommodate GIS and shaping GIS to suit the needs of the organisation.

The evolution of the organisational changes normally includes:

- development of a business concept and identification of goals
- appraisal of current:
  - tasks, work functions and routines
  - basic data
  - information products
  - information and data flow
  - data processing infrastructures
  - potential users
- review of the experience of others with GIS
- cost-benefit analyses, including delineation of:
  - assumptions for realising yields
  - measures to be enacted to realise benefits
- identification of what must be automated
- description of future data flow
- choice of strategies related to impetus, resources, assignment of priorities to data and uses, financing and organising
- data structuring
- development of a plan for further conduct, including project reorganisation

Current trends tend to regard information as a common and independent resource to be made available in numerous ways through the integration of data processing schemes. Consequently, planning should not be constrained to the organisation itself, but should include external users, data suppliers and other cooperating organisations. Hence several organisations may often participate in a single GIS project. Preparatory work should also aim to instill motivation, build competence and promote maturity within the organisation. Indeed, the organisational evolution should itself draw on the expertise of:

- professionals who know the tasks
- executives who know the organisation
- staff
- system specialists and/or external consultants

Evaluations of the various aspects of data processing and computer hardware and software should be aligned with activities relating to the evolution of organisational changes.

11.2.1   DEVELOPMENT OF A BUSINESS CONCEPT AND IDENTIFICATION OF GOALS

Many different motives may lie behind a decision to incorporate GIS into an organisation. Management may wish to increase internal productivity or external competitive ability. There may be a professional interest in new technologies. Or the organisation may be obliged to respond to internal or external pressures to approach tasks in a new or more vigorous way.

The motives for changing technologies must be thoroughly understood, for only then can goals be identified with sufficient clarity as to become attainable. This obvious requirement often means limiting the overall scope of the project.

Those parts of the organisation concerned with the change of technologies must also be identified in order to permit assignment of internal and external markets. The marketable entities consist of the GIS goods and services generated. These must, in turn, be defined and test marketed.

Initially, GIS must often be "sold" within an organisation and GIS "salesmen" - enthusiastic persons capable of promoting GIS - are sometimes needed to trigger the various processes leading to the implementation of a GIS facility.

11.2.2   APPRAISAL OF CURRENT TASKS, USERS, DATA AND DATA FLOW

One of the assumptions fundamental to the implementation of any new technology is that it performs old tasks in new ways, performs tasks that were previously extremely difficult or impossible, or offers a combination of these two advantages. One of the fundamental starting points for appraising a new technology therefore is clear identification and description of the current tasks. The appraisal should, of course, include all the users, regardless of whether the system to be implemented is limited to a single bureau or agency, or is to be a larger, multi-user system.

The first step of an appraisal, then, is to compile an overview of actual and potential users: the next is to elicit the relevant information from these users. Because questionnaires are always subject to misinterpretation and users are not always aware of their specific needs, user studies are best conducted through personal interviews. Unfortunately, these are time consuming and costly, and in most cases the survey must be limited to a representative sample which should, of course, include the major users who will benefit most from the new GIS technology. The sample should also include a broad spectrum of other users in order to give the best overall picture of what is required.

The user survey should identify:

- all production processes
- all files and filing systems
- descriptions of all data, including content, accuracy, completeness and currentness.
- all relevant details, such as data flow, documents and communications between processes, staff and internal and external files
- all data sources
- all data users
- how data are used:
  – tasks and products
  – decision making
- how data are managed and maintained
- descriptions of current methods, including any current computer systems
- key personnel

The depth of detail should, of course, suit the goals of the assessment. A detailed overview of a production process, for instance, is essential only if one of the purposes of the transfer to GIS is to rationalise production.

Basic data are best described in a "catalogue" describing:

- what the data represent, with a schematic division into logical levels, units (categories, groups etc.), and relations
- who produces the data
- data storage media and volumes
- data updating frequency

Product information is vital for several reasons. In some cases, production processes may depend on the information and organisational structures required to produce specific products. In other cases, product improvement or process optimisation may be important factors.

A clear diagram of information flow provides an overview of the various processes and their interrelationships. However, because data flow may be complex and therefore difficult to describe, special techniques, such as structured systems analysis, must be used. Assessments of current tasks must also contain evaluation of factors having financial impact, including:

- problems related to the conduct of such current tasks as:
  - proliferating delays
  - a decline in quality
- the ideal situation sought by individual users of geographic information, e.g.
  - the demand for new products and services
  - ways of changing current products and services
  - necessary organisational changes
- likely time or cost savings resulting from realisation of the ideal situation
- side effects of the new technology
- number of present and potential future users

Human factors, including the working environment, individual expertise and the like, should also be identified.

Many organisations are computerised and thus already have data processing infrastructures in place. As these infrastructures may affect the introduction of GIS facilities, they should therefore be evaluated with respect to:

- system approach and software systems
- user interfaces
- databases
- operating systems
- programming languages
- hardware
- communications facilities

11.2.3   REVIEW OF THE EXPERIENCE OF OTHERS WITH GIS

All organisations and activities have their own particular requirements. Nonetheless, a review of the experience of others is useful whenever new technology is implemented.

As is the case with all technologies, operating GIS facilities fall into three broad classes:

- "state of the art" or "leading edge": usually only a few organisations are at the forefront of the field because keeping abreast of the latest developments is expensive and requires expertise to deal with unproven approaches.
- proven: in the burgeoning field of GIS, the bulk of facilities now in operation comprise "off the shelf" hardware and software adapted to specific tasks and proven in operation. Directly or indirectly, the cost-benefit ratio of such a facility is usually near optimum.

- outmoded: ageing facilities of the "proven" category are frequently reliable producers. However, as constituent parts or whole facilities become outmoded by newer designs, cost-benefit ratios are less favourable because greater benefits may be realised through upgrading.

In most cases, the experience of organisations with "proven" facilities are the most valuable in designing a new GIS facility. "State of the art" facilities are usually too expensive or too exotic for organisations not directly involved in GIS research and development. "Outmoded" facilities are seldom of interest to an organisation contemplating a new technology.

Whenever there are few or no relevant users, for instance when GIS is to be employed in a totally new sector, studies must be conducted in two phases. Firstly, those aspects of the new application which may be similar to those of existing applications should be singled out. Secondly, those aspects should be evaluated, one-by-one, in order to form an impression of the mix of proven and new approaches in the new system.

A standard questionnaire should be used to compile information from varying sources and care should be taken in selecting the respondents if the information gleaned is to be meaningful. Experienced or early users often provide valuable insights, as do those who physically use a GIS facility in their daily work. In short, interviews with "hands-on" users are often more productive than those with an organisation's executives. Nevertheless, executives are more likely to provide a better overview of the organisational and financial impact of GIS. The query can be regarded as having been successful if it delineates the changes wrought by the initiation of GIS activities within the organisation.

One difficulty with such surveys is that even the most experienced users seldom fully exploit the potential of a GIS facility. The potential benefits of a system are evident only after its goals are defined. Consequently, user surveys can provide an overview that is valid only at the time they are conducted.

11.2.4   FINANCIAL EVALUATIONS: COST - BENEFIT ANALYSES

User surveys and pilot projects provide the bases for analyses. The conclusions from them should state the goals and identify the measures to be enacted. Analysis of cost-benefit ratios is an important part of the overall analysis.

The purpose of financial evaluations is mainly to:
- strengthen the basis for decision making
- increase benefits
- build awareness of financial impact
- develop motivation
- enhance planning

Other motives include:
- rank prospective projects (GIS vs. others)
- rank alternative designs and methods
- delineate the need for further studies or planning
- rework the project for greater profitability

Theoretical Basis

In business, costing tends to focus on efficiency and increased productivity resulting from a new technology is usually expressed in terms of profitability.

Governmental agencies evaluating changes to new technologies, however, often take socioeconomic factors into consideration. This mandates a broad view that includes both internal efficiency and externaleffectiveness. Socio-economic costing, therefore, is based on the inclusion of all parties likely to be affected by the change under consideration.

In a company or municipal department, inner efficiency, or benefit, can be measured either in terms of increased productivity or as savings in time and money. However, time savings are relevant only if they can be used to equal advantage elsewhere. From a socioeconomic viewpoint, such savings seldom pose problems when there is a shortage of qualified workers. During high unemployment, however, they may cause redundancies, thus increasing the number of unemployment benefit claims. Considerations like these dictate that for analytical purposes, a minimum tangible benefit level should be set.

The external effectiveness of an organisation is determined by how well it serves its users, both public and private. For users, increased effectiveness is manifested in lower costs, higher productivity or both.

Changing a technology often upgrades executive awareness, improves service and creates more meaningful work for the staff. Whilst such ancillary effects are difficult to quantify, they should nevertheless be included in an overall assessment.

Cost-benefit analyses customarily weigh various alternatives against each other with respect to their applicability over a prescribed time period. The zeroalternative, which retains the status quo with no change of technology, is almost always included. This requires a prediction of future events without the new technology. Other alternatives may incorporate varying degrees of technological change, but they must be pure; that is to say, each individual cost-benefit assessment must be independent of the costs and benefits that would accrue if there was no technological change.

In principle, socioeconomic evaluations and business economic evalu-ations comprise the same phases:

1. Problem identification
   - conceivable potentials of the new technology
   - identification of users
2. Pinpointing various courses of action
   - zero alternative (status quo) and technical alternatives
   - investment in the new technology vs. other investments
3. Listing and evaluating the cost-benefit factors of all alternatives
4. Reference benefits and costs to a single point in time for accurate comparison
5. Formulate suitable selection criteria so that consistent priorities may be assigned to the various alternatives for investment.
6. Monitor the outcome and verify results.

Costs

Commercial cost accounting usually is straightforward. Yet when socio-economic factors are involved, cost accounting becomes tenuous, because the complete effects of investments elude concise definition. For instance, a national map service may accurately estimate the initial cost of a new map series, as the costs of geodetic data, map making, field surveys, reproduction and the like are well known. However, the ultimate distribution of costs among the various end products, as ordinary map sheets, digital map data, and so on, is seldom as accurately known.

Cost accounts usually include:

- Inception studies
- Software
- Equipment maintenance
- Database creation
- Conversions
- Software development
- Supplies
- Administration
- Insurance
- Equipment
- Pilot project
- Software maintenance
- Communications/data network
- Training
- System operation
- Updating databases
- Cost of capital
- Rent

In addition, there may be intangible liabilities including:

- increased vulnerability due to reliance on computers, which may fail due to hardware or software malfunctions or failures
- poorer working environment, due to equipment-generated noise, tedious digitalisation tasks, etc.
- increased entry-level competence can bar some staff members

Benefits

The tangible benefits which may accrue can be divided into three categories:

- Benefits in resources
  - fewer tasks
  - less manual intervention
  - reduction in permanent staff required
  - reduction in temporary staff required
  - less overtime
  - savings in consultant fees
  - cheaper updating

- Benefits in products and services
  - information more rapidly available
  - greater task turnover
  - more rapid production
  - new products and services
  - faster processing
  - potential sales of products
  - more efficient invoicing of fees, etc.

- Benefits in output
  - fewer persons involved
  - greater turnover
  - improved internal and external communication
  - lower costs for others

The intangible benefits may include:

- Higher quality goods and services
- More conclusive decisions
- More rapid decisions
- More, better-used information and improved service
- Finer analyses
- Superior plans, strategic positioning and management
- Greater understanding of problems
- Enhanced expertise and more challenging work
- Stronger competitive ability
- Upgraded capability of organisation
- More career options
- Less routine work
- Keener financial management

All vital intangible effects, positive and negative, should be described and incorporated in the decision making process.

Cost-benefit analyses appraise costs and benefits in measurable terms, according to the prime rule of the relationship stated in Eq. 11-1.

benefits (or costs) = quantity X unit price   (11-1)

A typical computatation labour savings expressed in man-years, based on a productivity increase of 10% per staff member. The unit price would then be the average staff labour cost per year.

Costs and benefits often occur at different times, with costs usually being incurred before benefits accrue. Both costs and benefits may be divided into non-recurring values and annual values. These differences must be resolved so the various values may be directly compared. Future values must be stated in terms of present values, uncorrected for price level fluctuations. For instance, an income of one thousand monetary units received five years hence will, due to interest, be worth less than a present income of one thousand of the same monetary units. Consequently,future values must be discounted to permit assessment in terms of present values.

The usual procedure involves computing the Net Present Value (NPV) of a potential investment or project. The NPV is the aggregate difference between the discounted value of the expected benefits, or returns, and the discounted value of the expected costs. It is computed by the relationship of Eq. 11-2.

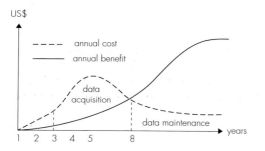

Fig. 11.4 - The benefit will come after a period of investment. Cost and benefit curves indicate that data acquisition and maintenance normally accounts for 60% to 80% of aggregate costs.

Year	Annual costs 1.000 US$		Annual benefit 1.000 US$	
	1991 US$	Present value (1991)	1991 US$	Present value (1991)
1991	200	200	50	50
1992	300	280	100	93
1993	300	262	200	175
1994	200	163	400	327
1995	200	153	700	534
1996	200	143	800	571
1997	200	134	800	535
SUM:	1.600	1.335	3.050	2.285

Cost	Benefit	Benefit/cost	Benefit ÷ cost
1.335	2.285	1.71 : 1	US$ 950

Fig. 11.5 - Typical computation of NPV, based on a discount rate of 7% and a time horizon of 5 years.

$$NVP = \sum_{t=0}^{T} \frac{B_t - C_t}{(1 + r)^t} \qquad (11\text{-}2)$$

where B is the benefit, or return, and C the cost at time t in years, r is the discount rate, and T is the time horizon, the point in time defined as the end of the economic life of the project or the time at which the assessment no longer evaluates costs and benefits. A typical computation is shown in Fig. 11.5.

The final computational result strongly depends on the interest rate chosen. In general, the interest rate should be high for high-risk projects, short-term projects, or combinations thereof. In Norway, the Ministry of Finance uses a rate of 7% in NPV and similar calculations. The rate used is often the difference between the rate of inflation and the borrowing rate.

The time horizon is determined by the projected economic life of the investment or project involved. The time horizon of most computerisation projects is five years. However, the time horizon of GIS projects is longer, usually about ten years. This is because the data employed in GIS are valid over longer periods of time.

Alternatives are often ranked in terms of their respective benefit-cost ratios, which are the ratios of the present values of their aggregate benefits to aggregate costs, as indicated in Eq. 11-3.

$$\text{Benefit-cost ratio} = \frac{\text{Net present value of benefits}}{\text{Net present value of costs}} \qquad (11\text{-}3)$$

The benefit-cost ratio usually provides the best index for ranking alternatives. However, in some cases, such as alternatives involving considerably different levels of costs are involved, the aggregate net benefit is also useful. As indicated in Eq. 11-4, it is the difference between benefits and costs:

$$\text{Net benefit} = \text{Net present value of benefits} \div \text{Net present value of costs} \qquad (11\text{-}4)$$

Methods which do not include considerations of time and interest rates can not be used as profitability criteria.

Sensitivity Analysis

The cost-benefit analyses involved in decision problems are invariably based on assumptions. The roles of the assumptions, however, is not always clear at the outset. Therefore, a sensitivity analysis is often performed to identify the assumptions to which the outcome of the cost-benefit analysis is sensitive.

A sensitivity analysis involves changing the assumptions or parameters of the decision problem to reveal how the changes affect the outcome. The analysis involves individual computations, which may be based on pessimistic or optimistic assumptions. The parameters which may be changed include:

- Time horizon
- Discount rate
- Order of activities
- Rate of investment

Evaluating Alternatives

There are no exact measures for assessing cost-benefit alternatives. However, rule-of-thumb experience indicates the following:

1. Measures which result in a benefit-to-cost ratio of 2:1 or more should be enacted, both when assessed in commercial terms and in socioeconomic terms.
2. Measures which result in a benefit-to-cost ratio between 0.8:1 and 2:1 should be analysed further with an eye to cutting costs or increasing benefits.
3. Measures which result in a benefit-to-cost ratio of less than 0.8:1 should be abandoned.

11.2.5   DEVELOPING A STRATEGIC PLAN

The pre-project phase should result in a strategic plan for the activities involved in initiating GIS. The plan should contain:

1. Recommendations for a pilot project, if any.
2. Descriptions of future activities.
3. Ranking of functions to be automated.
4. Statements of the assumptions requisite to benefits and identification of the organisational and technical measures and decisions required.
5. Time plan including schedules and milestones for the various activities.
6. Statement of staffing needs.
7. Investment plan and budget.
8. Requirements for expertise and training.
9. Restructuring of the organisation.
10. Financial plan.
11. Project organisation.

The technologies involved in GIS are developing rapidly and continuously. So GIS users may improvise new applications at any time. Consequently, the strategic plan should be sufficiently flexible to allow its conduct to be

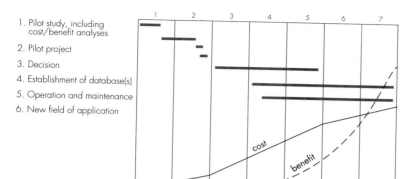

1. Pilot study, including cost/benefit analyses
2. Pilot project
3. Decision
4. Establishment of database(s)
5. Operation and maintenance
6. New field of application

Fig. 11.6 - Idealised initiation of GIS, where costs and benefits accrue during the project.

influenced by prevailing conditions.

The ranking of alternatives in the plan is based on the results of the cost-benefit analyses, in which investments in GIS are tested against imminent or current investments elsewhere in the organisation. In addition, public agencies must consider their official duties.

With a suitable strategy, the optimum initiation of GIS activities might be viewed as shown in Fig. 11.6.

A strategy plan should identify and illuminate:
- What is to be automated
- Future data flow
- Start date(s)
- Level of investment
- Geographical coverage
- Organisational matters
- Budget and financing
- Project organisation

What is to be automated

Current data flow, production lines, products and cost- benefit analyses form the basis for the macro view of what is to be automated. The macro views may include:

- Which work processes to be replaced by GIS
- Which files are to be replaced by GIS databases

Which lines of communication are to be replaced by electronic communications via data networks.

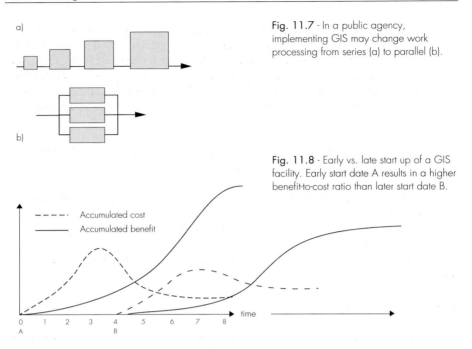

Fig. 11.7 - In a public agency, implementing GIS may change work processing from series (a) to parallel (b).

Fig. 11.8 - Early vs. late start up of a GIS facility. Early start date A results in a higher benefit-to-cost ratio than later start date B.

Future Data Flow

The description of future data flow must relate to the functions to be performed. Comparing future data flow to current data flow helps identify which work functions will become superfluous, which must be added and which should be modified.

In a public agency, for instance, the change of overall data flow caused by a change from conventional means to GIS may be regarded as being a change from series to parallel processing of work, as illustrated in Fig. 11.7.

Start Date(s)

One of the prime decisions to be made in all changes of technology is whether to start now or later. The KVANTIF study conducted in Scandinavia indicates that no users with operating GIS facilities felt that they should have postponed their changes to GIS. If fact,some felt they should have started earlier. The only regret voiced was that some felt that they might have started in another way. In short, the study indicated that an organisation contemplating GIS should initiate the relevant activities as soon as possible.

In general, the later the start-up, the greater the cost of establishing databases and the longer the time before benefits are realised. As illustrated in Fig. 11.8, a late start may cost no more than an early start, but may result in lesser aggregate benefits. The benefit-to-cost ratio is then less.

One of the more common justifications for postponing an investment in a new technology is that new and better technology will soon be available. The experience of GIS users thus far indicates such postponements seldom prove profitable if the system is implemented in a proper way. The continuous evolution of all the technologies involved in GIS implies that the basic decision of whether to invest now or later remains, no matter at what time it is made.

In summary, the factors that should influence the decision of when to start include:

- The relevant technologies are available, so they no longer comprise a bottleneck.
- Postponing a start postpones benefits and thereby lessens project profitability.
- All technologies now evolve continuously, so there is no one best time to start.

Level of Investment

The crucial choice to be made in investment is usually between constraining investments in new technologies within conventional budgets and considering more extensive investments.

As is the case for almost all changes from traditional to electronic technologies, the initial investments are relatively large yet short term. This trend should be exploited, as by establishing GIS databases quickly to realise the benefits they generate. Prolonging the establishment of a database, as by stretching it over phases involving both conventional analogue and computer digital data, only escalates costs without correspondingly increasing benefits. As shown in Fig. 11.9, cautious investment often yields less than aggressive investment.

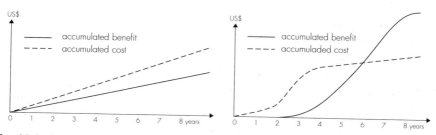

Fig. 11.9 - Aggressive investment often yields more than cautious investment. As shown in a), linear investment within the constraints of a conventional budget produces linear benefits, which may prove unprofitable. As shown in b), a major early investment, such as in the conversion of an entire database, can yield benefits that rapidly exceed costs, for greater profitability.

Once a GIS facility is planned, the question arises as to which part of it should be installed first. From an investment viewpoint, the part that yields the greatest proven benefit should be first. But for most users, a part comprising several georeferenced themes concerning a single area or region provides the more comprehensive initial benefit. This is because the various themes are often mutually synergistic. For instance, a digitised property register enables GIS to provide information on properties, and a digitised road data register enables it to provide information concerning roads. However, if both digitised registers are available, the GIS can provide additional information. It can then provide answers to queries as "which properties front onto which roads," "which properties are affected by a proposed rerouting of a road," and the like.

In conclusion, the guidelines for levels of investment may be summarised:

- Think big but start prudently.
- Make major investments after a pilot project phase.
- In switching from conventional means to GIS, major investment should be committed in the initial implementation phase in order that the project be profitable.
- Data should be rapidly converted to digital form.
- Several themes pertaining to a region or area can synergistically increase benefits to users.

Geographical Coverage

As a rule, the basic decision to be made concerning geographical coverage within the scope of a GIS is whether to provide full digital coverage of all regions and areas involved or fully digitise information only for areas of greater activity and more rapid change. Again, the economics of the relevant benefit-to-cost ratios are the best criteria.

Different user organisations have differing needs for the geographical coverage of digital information. The distinctions involved often are dictated by functional or administrative divisions, as townships,regions, counties or entire countries. In general, a need for digital geographic coverage must be fulfilled completely before the user realises benefits. This is because stage-wise shifts invariably include periods when both conventional and digital data are used, which usually increases the work involved in producing a given final product.

Consequently, as a rule the most cost-effective approach is for an organisation to start with full geographic coverage, even though full coverage requires a greater initial investment in data acquisition, operation and maintenance. The distinction between the coverage needs of various organisations is then not one of coverage itself, but rather of the regions or areas of

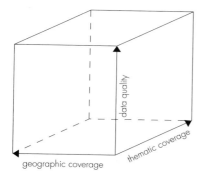

Fig. 11.10 - Usefulness depends on geographic coverage, thematic coverage and data quality. These three parameters may be viewed as defining a "usefulness cube." (Ken Jones)

responsibility involved. A managerial organisation, as a major governmental agency, usually must maintain full geographic coverage of its entire area of responsibility. An organisation dealing with projects, as a road works department, need only cover involved in its projects.

In conclusion, the guidelines for geographic coverage may be summarised:

- At least one theme covering an entire region or area of responsibility must be converted before benefits may be realised.
- Stage-wise conversion to GIS, in which conventional and digital data may be used simultaneously, should be avoided.
- Managerial organisations have the greatest needs for coverages of entire regions or areas, while organisations operating on a project basis have lesser needs.

The potential usefulness, or benefit accrued depends to a great degree on geographic coverage, thematic coverage and data quality. These three parameters may be regarded as a "usefulness space" or "usefulness cube," as shown in Fig. 11.10.

Organisation matters

The efficient exploitation of a new technology in an organisation often mandates alterations in work routines and chains of command which, in turn, affect the overall organisation. In practice, altering an organisation may prove difficult, both because the organisational structure is intangible and hence difficult to define, and because there invariably are both formal and informal positions in all chains of command. Changing the organisation changes staff authorities and relationships. And staff changes always bring in human factors that are difficult to predict or control.

Consequently, organisational matters are vital in all initial implementations of GIS facilities. The organisational problems are often more complex and more crucial to success than are the technical problems involved. As a rule, technical problems can be solved in a straightforward manner, by pur-

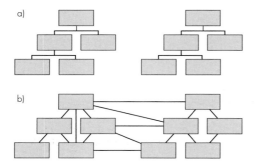

Fig. 11.11 - Transition from a hierarchical organisation (a) to a network organisation (b).

chasing and installing new equipment, new software modules, and so on. Purchasing incurs costs, but in a well-planned project, these costs are anticipated. Changing and replacing staff members is less straightforward, and may trigger unanticipated problems. Hence, organisational matters usually require more continuous management attention than do technical problems.

Consequently, one goal of any inception study should be to delineate alternative future organisation models and to recommend ways of testing their validity. New staff positions must be described, complete with descriptions of tasks, duties and responsibilities. In principle, new tasks and new data flows may be described independently of the persons, groups or departments ultimately responsible. So new staff structures must be defined, and management modified accordingly.

Information availability is vital to a viable GIS facility. Whenever information availability is restricted, GIS utility suffers. So the common penchant of an agency or department to monopolise its own information is one of the major foes of successful GIS. Consequently, one of the problems encountered in implementing a GIS facility may involve combating the bureaucracy that blocks information flow.

Other subjective factors can further complicate the initiation of a GIS facility. Human habit apparently dictates that about a quarter of the personnel in any organisation always prefer status quo and will oppose any change whatsoever. Except in high-tech firms, executives are often indifferent to newer information technologies, partly out of ignorance and partly from being overly concerned with cutting costs rather than increasing benefits. However, as executive support is often crucial to the success of a GIS project, initialpresentations may be more readily received if they contain clear executive summaries and concise examples of the utility of GIS in sectors known to be of interest to the executives concerned.

Successful projects are often enthusiastically conducted by middle management, who often seek the benefits of new technologies while resisting extensive organisational changes.

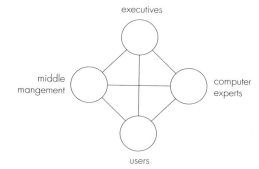

executives

middle
mangement

computer
experts

users

Fig. 11.12 - To be successful, a new technology must be accepted at all levels in an organisation.

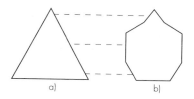

a)   b)

Fig. 11.13 - The introduction of a new technology in an organisation often changes its power pyramid from the traditional triangle (a) to a barrel profile (b) in which fewer employees are involved in routine tasks and middle managers assume more comprehensive roles, both upward and downward from their locations in the traditional pyramid.

Specialists are essential in the execution of a GIS project, but seldom are capable of addressing the myriad of detail involved in a project. For instance, computer experts are usually so engrossed in their disciplines that they lack the broader view requisite to dealing with organisational problems. Operators and other direct users are similarly constrained within the frameworks of their jobs.

As shown in Fig. 11.13, the introduction of a new technology in an organisation often changes its power pyramid.

As discussed above, the needs for expertise are often underestimated. Without the requisite expertise, progress can be frustratingly lethargic. Consequently, maintaining and building staff expertise should be a top priority task. Personnel management for a transition to a new technology should include:

- identification of factors affecting personnel
- a plan for training, retraining and relocation
- concise job descriptions, including responsibilities
- employee involvement in planning
- clear definitions of positions and salaries
- a plan for job rotation
- delineation of the simplest possible organisational structure, including:
  - management authority
  - delegation of responsibility
  - division of work

- clear communication to and from executives, between organisations and within individual organisations
- retain the organisation's advantages and strengthen:
  - team work
  - goal-orientation and adaptability
  - identification with the project
- suit staff proficiencies and size to the tasks involved
- create an environment of challenging work and career advancement

The most important organisational aspects may be summarised as follows.

- Executive involvement is crucial to success.
- Organisational problems are usually greater than technical problems.
- Organising or reorganising should hinder the monopolisation of information.
- The introduction of GIS effects changes in existing routines for information interchange between and within individual units of an organisation.
- Altered work routines mandate organisational changes.
- At least a quarter of the personnel of an organisation can be expected to oppose change.
- Operator cooperation must be enlisted.
- The initial stages of implementing a GIS facility may be project oriented. The organisational alterations should be tested before being finally enacted.
- Long-term organisational changes may be made after the initial operational phase of a new GIS facility.

Budget and Financing

As is customary, a GIS implementation project should have a budget that includes distributions of activities over the duration of the project and permits monitoring.

Projects based on meticulous cost-benefit analyses are usually the most viable. In many cases, there is not just one project, but several, as several organisations may jointly be involved in establishing a GIS facility. In these cases, costs are usually apportioned among the organisations according to benefit accrued. However, costs of facilities used by many organisations may be met in other ways, as by charging annual fees, assessing an initial outlay plus an annual fee, paying according to ability, and so on.

Governmental agencies often must decide whether to finance a new facility through their normal fiscal allocations or by charging use fees at market rates. As discussed later in this in chapter, the problem is central in the organisation of national databases.

Project Organisation

In practice, GIS implementation projects are often understaffed. Therefore, at the outset, staff requirements should be carefully assessed, and the staff identified should be made available to the project. As a rule, this usually requires either that staff members be relieved of their customary tasks and assigned to the project, or that personnel be hired specifically for the project.

A strong, independent project organisation is essential to the success of any major GIS implementation. The project need not be permanent, but can be disbanded when the GIS facility becomes operational. GIS, computer and other experts are usually more readily hired for a project than trained for the tasks involved. All projects need enthusiasts and problem-solvers, who need not be specialists and who can advantageously be recruited from among the users of the information products involved.

External consultants often are most beneficial in strategic sectors of a project, as they are not involved in the organisation and hence frequently can resolve difficulties and dispute more easily than can members of the organisation staff.

11.2.6   DEVELOPING A LOGICAL DATA MODEL

In the computer sciences, data modelling is regarded to be part of the development of a data processing system. This view is common, but not always correct. All processed data must have initial sources and final uses and must relate to the real world outside the computer environment. A good data model is part of the specification of a data processing system, not the result of it. Therefore, a data model should be devised by professionals expert in the field of application involved.

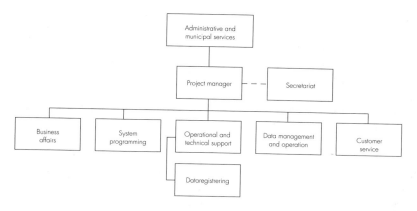

Fig. 11.14 - Typical organisation of a GIS implementation project (Burnaby, Canada).

Data models and data structures are discussed in detail in Chapter 3 and 6. In summary, data models comprise objects, specified by five parameters:

- object type
- geometry
- attributes
- relations
- quality

These parameters should be described in a data dictionary.

The procedure for data structuring includes several topics previously discussed in Chapter 6:

1. Defining applications
2. Stipulating constraints, tasks and requirements
3. Ranking according to benefit-to-cost ratios
4. Descriptions of attributes
5. Coordination of conditions and definitions
6. Iteration of items 1 through 5 above, if necessary
7. Choice of geometric representation: raster or vector, points, lines and surfaces
8. Descriptions of relations between objects
9. Quality requirements based on cost-benefit analysis: geometric accuracy, updating frequency, completeness and resolution, geographic coverage, logical consistency
10. Coding of data with respect to identity, object types and attribute values

11.3 IMPLEMENTING A MAJOR GIS FACILITY: SYSTEM DEVELOPMENT

The development of a GIS system pertains to all aspects of the computerisation and data processing involved, and may include:

- design requirements
- pilot project
- choice of system
- system implementation
- database design
- creating databases
- operation and maintenance
- evolving new applications

A GIS system is usually developed partly in parallel with the relevant organisational activities. For instance, design requirements may be compiled when the organisational strategy and future data flow are defined. The choice of a system or supplier may be made after the benefit-to-cost ratios have been finalised. Work routines may be delineated as the system nears completion.

11.3.1   DESIGN REQUIREMENTS AND SYSTEM SPECIFICATIONS

The requirements pertaining to the various tasks and routines, sources and models, and components and sub- systems comprising a GIS facility are usually incorporated in invitations to bid issued to selected suppliers. In most cases, these design requirements both reflect the results of user surveys, cost-benefit analyses and strategic planning, and involve compromises concerning what is current available on the market. Design requirements may be compiled for:

- organisation(s) evolved
- tasks automated
- operations conducted
- products produced
- data employed:
  - type
  - amount
  - quality
- data sources
- data model(s)
- time horizon
- training
- standards and their validity
- documentation
- data processing infrastructures in place

Hence, the design requirements of a system stipulate how it is to function, not how it is to be implemented. In turn, specifications are compiled for the equipment, sub-systems, routines and other specifics of implementation that provide the various functions. At this stage, responsibility for the implementation phase of the project may be defined. The assignment of responsibility varies considerably, depending on the organisation involved and on the nature of the markets that supply goods and services for the implementation. In general, an organisation may elect to retain or assign responsibility for a system.

Retain system responsibility: The organisation, or a consultant it engages, detail designs the system and issues specifications for equipment, sub-systems and the like. The organisation itself is then ultimately responsible for the system(s) being able to fulfil the design requirements.

Assign system responsibility: The organisation issues its design require-ments and invites suppliers to respond with bids that include the detailed specifications of how the requirements are met. The supplier(s) selected carry the ultimate responsibility for the system(s) being able to fulfil the design requirements.

There are many mixes of these two main approaches. Organisations com-petent in the fields involved may prefer to issue specifications, yet assign system or sub-system responsibilities. An overall system may be split up into sub-systems, some with supplier sub-system responsibility and some without.

The basic decisions of who performs detail design and compiles specifi-cations and who assumes system or sub-system responsibility invariably involve tradeoffs between in-house and external activities.

11.3.2   PILOT PROJECT

A pilot project entails an experimental small-scale implementation in which techniques, equipment and sub- systems for use in a full-scale GIS are tested in advance. Pilot projects should:

- provide better bases for choice of system(s)
- test various production methods
- contribute to verification of cost-benefit analyses:
  - better time estimates for data acquisition, etc.
  - better assessments of time gains, etc.
- identify system faults
- assess organisational problems
- assess the staffing requirements for and time consumed in transferring data from conventional to digital form
- help develop the organisation:
  - increase appreciation of the GIS concept
  - mediate organisational problems

The scope of a pilot project depends on the complexity of the organi-sation(s) implementing the GIS facility. From a practical viewpoint, a pilot project should be budgeted in both time and money. The equipment and sub-systems used should be as simple as possible, both to minimise loss should the chosen approach prove unsuitable and to ease the requirements for staff training and maintenance. As a rule, pilot projects should not take longer than one to two years.

Staff involvement in a pilot project should be restricted to those most concerned with the GIS facility. Although widespread familiarity with GIS technologies and with the pilot project is advantageous, project progress may be complicated if an excessivenumber of conflicting interests are involved.

Normally, one or more limited geographic test areas should be chosen. Ideally, the test areas should be dissimilar, as one of high complexity and one of relatively simple content.

The basic goals of a pilot project, to test approaches and prove viability, dictate maximum use of the experimental system(s) involved. Staff should accordingly be made available, least the pilot project be slowed for lack of manpower.

11.3.3   PILOT PROJECT OUTCOME AND THE DECISION TO CHANGE TECHNOLOGIES

If the outcome of a pilot project countermands GIS implementation, the pilot phase may be extended, different approaches may be taken, or the GIS project may be abandoned completely.

Before any of these steps are taken, the reasons underlying the negative outcome should be examined. As is often the case in assessments of new technologies, a GIS pilot project can flounder if its initial expectations are unduly high. Whenever suppositions of technical, financial or organisational benefit outstrip realistic goals, downfall and disappointment invariably result. These effects can scuttle even the best of projects. Hence, at the outset, prudent goals for a pilot project are preferable.

If the outcome of a pilot project supports GIS implementation, work should commence as soon as possible, particularly in choosing a system and starting data conversion activities. With a rapid decision and no dead time, the inertia of enthusiasm in the pilot project can carry over into the implementation phase.

11.3.4   EVALUATING AND CHOOSING SYSTEMS

Choosing a system is at once easy and difficult. It is easy because it deals with comprehensible details, as characteristics, prices, service and the like. It is difficult because the future uses of a system may be nebulous and because the drawbacks and disadvantages of the sub-systems considered are seldom obvious at the outset.

As in other disciplines, computerisation efforts in the geographical disciplines often overemphasise equipment and software. Regrettably, data, the greatest resource and value of a GIS facility, often seems of lesser interest.

The fallacy of focusing system concern on hardware and software to the exclusion of data is illustrated clearly by the longevities involved in a typical GIS facility:

Part of GIS facility	Average useful lifetime
• computer equipment:	2 to 5 years
• software programs:	3 to 6 years
• data:	15 to 20 years or more

The shorter expected useful lifetime of computer hardware and software combined with the lesser investment involved - from 10% to 30% of the aggregate cost of a GIS implementation project - indicate that the major concern is best directed elsewhere, to the data which are the backbone of an operating GIS.

The system specifications, which are based on the design requirements, include stipulations of:

- system solution
  - computer: capacity and speed
  - peripheral devices: number and types
  - communication
- software
  - functionalism
  - user friendliness
  - accessory modules
  - customizing
- maintenance
- training
- costs

There is no one ideal set of specifications. Organisations combine specifications in varying degrees to suit their individual needs. For instance, organisations with steadfast products and administrative tasks, as utilities, national map services, municipal authorities, and the like, usually emphasise stability and service, and choose systems accordingly. Agencies responsible for planning and resource management may favour greater flexibility and diversified applications. Schools and Universities inevitably favour elementary, inexpensive systems.

Several objective artifices may be employed to simplify choices and aid decisions. In all cases, requirements should reflect needs, not system parameters.

For instance, as illustrated in Fig. 11.15, systems may be numerically ranked by assigning weighted values to all parameters. Parameters that cannot be quantified may be assessed in "yes-no" terms, i.e. fulfils or does not fulfil a requirement.

Selection criterion	weight	rank				weighted rank			
		a	b	c	d	a	b	c	d
System cost	5	7	7	6	9	35	35	30	45
Database design	4	6	7	8	7	24	28	32	28
Man-machine interface	5	8	6	7	6	40	30	35	30
Equipment functionality	3	7	8	7	7	21	24	21	21
Own programming	3	7	5	5	6	21	15	15	18
Follow-up	4	7	6	6	8	28	24	24	32
Data interchange	5	8	7	6	5	40	35	30	35
Expansion possibilities	4	7	8	8	8	28	32	32	32
Documentation	4	7	6	6	5	28	24	24	20
Maintenance costs	4	9	5	6	6	36	20	24	24
Drawing functions	2	7	9	9	6	14	18	18	12
Unweighted sum		80	74	74	73				
Weighted sum		315	225	255	287				
Rank		1	3	4	2				

Fig. 11.15 - Typical comparison of four GIS, a, b, c and d, using initial ranking of 1 through 10 and weightings of 1 through 5.

The performance of a system as installed is vital, as it provides an index to payoff potential. The performance of a system at some future date is of lesser immediate interest. Performance may be assessed in terms of how a system executes specific functions, how rapidly it works, how simple it is for users, and how flexible is it in current applications? Such questions are best answered through testing.

Different systems are best compared by running specific test programs using uniform test data. This "benchmark" approach, or functional test, is valuable in assessing various systems and approaches.

The parameters to be considered in selecting and installing equipment include:

Installation details:
• floor and other structural loads
• furnishings
• mains power supply
• ambient temperature and humidity
• noise environment

Computer characteristics:
• computational capacity
• computational accuracy
• communication
  – local area network
  – data communications protocols
  – speeds
• number of peripherals supported
  – graphic terminals
  – alphanumeric terminals
  – printers

– plotters
– digitisers
- expansions accommodated
- compatibility

In practice, equipment use is often impeded first by computing capacity and the number of input-output interfaces available. The more important characteristics of the peripheral units include:

Graphic displays
- screen size
- image quality
- colours
- national character sets
- ergonomics

Plotters
- accuracy
- resolution (raster plotters)
- output format
- plotting speed
- plotting medium
- implements

Digitisers
- accuracy
- resolution
- size

The characteristics of tape stations, internal storage facilities, disk drives, alphanumeric terminals, printers, local area networks supported, data communications protocols available, etc., should be listed and evaluated in a similar manner.

The software requirements should include:

- structure of basic system and application programs
- system openness to support user-defined application programs, communications programs, etc.
- macro programming capability
- user instructions and system support documentation
- freedom to change databases
- facilities for restricted access
- operational records and user accounting
- tools for restructuring databases

- selection of languages available for entry of error messages and commands
- programming language
- support of data interchange with other systems
- man-machine interface (response time, error messages, commands)
- support and follow-up

The above list may, of course, be much longer, depending on the software application involved.

In general, the criteria involved in choosing a system depend on the nature and goals of the organisation involved. A case typical in GIS applications is the implementation of a joint use facility shown to have a benefit-to-cost ration of 4:1. The guidelines for its specifications are:

- start with a general-purpose system
- note that general systems generally entail more initial effort
- modular systems are easily expanded
- tailor-made systems constrain further developmentand limit the utility of databases
- tailor-made systems are rapidly obsolete and often expensive to maintain
- general-purpose systems ease information interchange and inter-organisational communication, which enhances benefit-to-cost ratios
- specialised systems can limit benefit-to-cost ratios when information is digitised
- evaluation should be as structured and as objective as possible

From the early days of computer systems up to the explosive expansion of computerisation triggered by the introduction of personal computers (PCs) in the early 1980s, systems were inevitably proprietary. Hence the choice of a system was tantamount to choosing a supplier. Vestiges of this trend remain, as many systems remain proprietary. However, internationalisation and inter-company tie-ups in the computer industry, trends towards open computer architectures, increasing inter-program compatibility and, of course, cloning and competition have all contributed to increasing the selection of suppliers capable of delivering a system or sub-system. The choice among competing suppliers usually includes considerations of:

- capabilities and service
- market position
- company stability
- price policy
- references

11.3.5   DATABASE DESIGN

A thorough database design is both essential to realising the functions required and for determining:

- scope, so only relevant data are entered
- ease of access to data
- efficiency with which functions are executed
- facility with which data are updated
- receptiveness to new types of data
- ease of restructuring data in the database

As discussed in Chapter 3 and 6, a data model provides the basis for a database design. The detailed design of a database depends on the system software and hardware chosen. In joint-use facilities, databases must be further designed for individual organisation data or for data common to all organisations involved.

The best approach is for each organisation to design databases that best suit its specific needs, and to incorporate direct or indirect inter-organisation communication via a joint database. Such a distributed approach requires that distinctions be made between joint data common to all organisations and individual data belonging to the individual organisations. The various individual and joint databases are then connected by a data communications network. The physical separation between the databases and the communication among them are not obvious to users, who access the whole. The distributed database approach also requires good updating routines and extensive standardisation.

DATABASE DESIGN

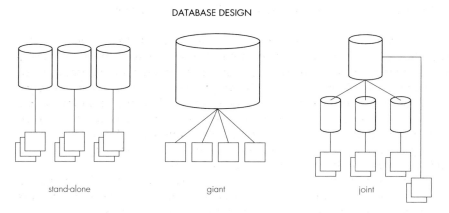

stand-alone   giant   joint

Fig. 11.16 - Basic database design may be divided into three categories, stand-alone, giant and joint. The choice between the three depends on design goals. The joint database design is the most complex, but provides the greatest benefit.

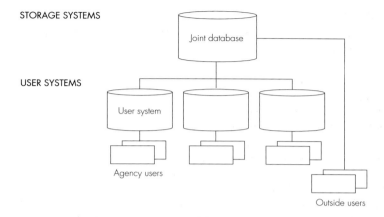

STORAGE SYSTEMS

USER SYSTEMS

Joint database

User system

Agency users

Outside users

Fig. 11.17 – In a joint database, some information is available to all users, while parts of the total information are stored in user databases. The joint database is usually separate and may be networked. Different types of user databases may be used. Outside users may be permitted to retrieve and present data.

As the number of users having on-line access to a database or inter-connected databases increases, the need to restrict access to certain parts of the data may also increase, for security, strategic or maintenance reasons. Such considerations must also be included in database design.

Software Dependent Database Design

In most GIS database systems, data are organised in geographic units, as map sheets, administrative areas or other delineated areas. The sizes of the areas involved may vary considerably within a single database. However, from the user viewpoint, the data involved appear "seamless." That is, the user does not "see" the borders between adjoining areas/sheets.

The size and shape of the storage units affect datastorage efficiency and accessibility and the ease with which data may be updated. Fixed unit size is most efficient whenever data are spread evenly over the entire region covered by a database. Small units are an advantage whenever data density is large. The scale of presentation and the sizes of areas which may be simultaneously processed also influence the choice of unit size. In practice, units are often chosen to correspond to the map sheets of the region covered by a database.

The geographic storage structure should be chosen with care, as established database structures are difficult to modify.

In most GIS, data are also organised in thematic layers, which correspond to the overlays of traditional map production. The choice of topics for the thematic layers determines how efficiently their data may be used.

Each thematic layer comprises types of objects that are to be processed together, so data use determines the allocation of objects to layers. For instance, one application may benefit most from having all linear objects, as roads, rivers and coast lines, in a single thematic layer, while another application might separate rivers and coast lines into separate thematic layers.

The greater the variety of uses of the data inthematic layers, the greater the number of thematic layers. However, data may be collected in a common thematic layer if they are:

- always used together
- logically related
- updated simultaneously
- valid for several functions, as lines indicating roads, property boundaries and land use boundaries
- used together in presentation, as all 0.3 mm wide lines, all areas coloured red, all place names, etc.

The uses of a database may change with time, so databases should be designed to be modified to incorporate new object types and attribute types, as well as to be enlarged.

Hardware Dependent Database Design

Database design depends on computer hardware with respect to:

- data volume
  - storage medium
  - storage capacity
- accessibility and speed
  - single or multiple users
  - number of work stations
  - on-line
  - off-line
  - storage medium
  - network

Different systems employ different approaches to data storage, and storage methods often may be modified or expanded. For instance, at the time of writing (early 1992), the storage capacity of stand-alone PCs is limited by the storage capacities of hard disks. Future memory capacities may be considerably greater, as optical disks and other storage media are employed in PCs. Even with current technology, effective PC storage capacity may be increased considerably by connecting a PC to a local-area network that has a larger server with greater storage capacity.

In many cases, the constraints imposed by data storage facilities are more administrative than technical, as organisations implementing new systems frequently wish to employ existing data processing infrastructure.

11.3.6   CREATING A DATABASE

Creating a database is both costly and time consuming, as well as demanding of facilities and staff until the database is finished.

As discussed above, the databases employed in a GIS facility should be compiled as rapidly as possible, in order that the data they contain be accessible as soon as possible. Consequently, both private firms and public agencies often arrange data conversion and entry work in round-the-clock shifts.

As they know their own data and professional fields best, most organisations elect to convert and enter data in-house. As the work load involved may be enormous, outside temporary staff are often hired for the routine tasks.

Identified user needs, preferably ranked according to benefit-to-cost ratios, should determine the data entered in a database. Clearly, data accessed by many users have the greatest utility. Therefore, database creation should start with joint data. At the bottom of the data ranking list may be information entered on the premise that it may be useful at some time. In many cases, if its use is uncertain, then it need not be entered.

11.3.7   SYSTEM OPERATION AND MAINTENANCE

Data must be maintained to retain their value. Normally, data maintenance in a major facility is considered only after the facility is fully operational, some three to seven years after the first data were entered. However, data maintenance is vital, from the very start.

Persons active in data acquisition are usually the most qualified to maintain the same data. Surveyors can both acquire and then maintain utility network data, planners can compile and then maintain zoning plans, and so on. The most direct approach is for the responsible professionals involved to maintain their own data via work stations accessing a joint database.

Maintenance routines should include deletion of duplicate entries, which are common in conventional manual systems. This implies that the various organisations accessing a joint database are obliged to depend more on each other. Consequently, all organisations involved must agree upon the maintenance routines to be used.

11.3.8   EVALUATING NEW APPLICATIONS

With time, most GIS users discover new needs and therefore new applications for systems and data. In many major GIS facilities that have been operational forsome time, users report as many as 40 new applications implemented after the system became operational. The new applications often differ considerably from those originally anticipated for the facility, and therefore are among the factors that increase the benefits GIS provides.

The procedure for assessing new applications is straightforward. After a GIS facility has been in operation for a few years, a survey of user needs may be made to chart new needs. New needs identified may be evaluated in terms of benefit-to-cost ratios, just as are the initial needs for which the system was originally designed. Likewise, the needs for organisational alterations should be evaluated, and alterations should be enacted using the same criteria as used initially.

Most new applications require programming work, and the ease with which they may be realised depends on the flexibility of the system implemented.

11.4   GIS AS AN ANCILLARY

As discussed in the first section of this chapter, GIS facilities may be divided into two groups, according to whether or not they affect the overall organisations. There are, of course, many borderline cases. But ingeneral, many smaller organisations, as smaller townships, utilities or agencies, may implement smaller GIS facilities for special and often internal purposes. From the organisational viewpoint, such smaller facilities may be regarded as ancillaries.

Nonetheless, the guidelines for assessing, planning and implementing a GIS facility are similar to those discussed thus far for major GIS facilities having far- reaching organisational impact. The differences involved are more of magnitude. The tasks involved usually are more easily defined. The total economic risk involved is usually more modest. Data flow is usually more readily delineated. And, as the GIS facility is ancillary to the activities of the organisation, the organisation can adapt to it more rapidly.

Many ancillary GIS facilities are project oriented. Typical facilities include digital terrain models used in planning residential developments, automation of township property registers, and so on. In most cases, the applications are clearly defined and the design challenge reduces to finding the most suitable system for the tasks involved. In practice, the system choices are often among various PC systems available.

The selection strategy can be reduced an elementary "four S" guideline:

- small
- sure
- seen
- success

The initial system should be small to minimise financial risk and to speed familiarisation with the new technology.

The organisation should be sure that the tasks identified are clearly defined, necessary and sufficiently limited that they may be easily realised.

The results should be seen soon, both to justify investment and to encourage the staff involved.

When these three requirements have been met, a measure of success will be at hand, relatively rapidly at a minimum investment.

11.5   CREATING NATIONAL GEOGRAPHIC DATABASES AND DEVELOPING NEW BUSINESS SECTORS

Creating a national database raises numerous strategic questions, as public contra private financing, product organisation, data content and quality, and so on. Many countries have addressed these questions and have created national databases a variety of data, including:

- property registers
- road data
- environmental data
- elevation data
- statistical data
- addresses
- building data

The creation of a national database may be compared to developing a new market. So the guidelines are equally valid in the public and private sectors.

11.5.1   PRINCIPLES OF EVOLVING STRATEGIES

All strategies start with the basic initial decision of whether to maintain status quo (the "zero alternative") or to select among alternatives to create a national database. Of course, the status quo, or zero- alternative, decision requires no strategy. Strategies afford means of selecting among alternative courses of action.

Strategies are best based on cost-benefit analyses, which rely on costs and benefits being expressed in monetary units. Quantifying the costs and benefits of various courses of action is perhaps the greatest challenge of the entire strategic decision process.

Strategies must also include considerations of organisational factors, as production, storage, distribution and updating.

All planning presupposes some experience with the subject. However, as a national geographic database is often the first of its kind in a country, the relevant experience usually is sparse. Consequently, the evolution of a strategy usually includes activities dedicated to defining the issues at hand. In general, a strategic plan may evolve in five phases:

1. Identification of current status, or zero- alternative: the first user study
   – among users
   – among providers (including own organisation)
2. Elaboration of a product concept: defining offerings:
   – which data users consider vital
   – characteristics of data offered, as accuracy, currentness, etc.
   – initial matrix of products offered, as in three categories of increasingly finer service
3. Formation of a cost model
   – that simulates costs of the various product alternatives
   – that ascertains product cost chains
4. Delineation of and conduct of market analyses: the second user study, which aims to identify:
   – user ranking of products and inclinations to pay for products in the various application groups
   – size and potential of the sundry market segments, including factors that may detract from market development
5. Devising a model, based on the cost and market analyses, that promotes identification of an optimum product concept

11.5.2 MARKET ANALYSIS: FIRST USER STUDY

The first step of the market analysis is to detail how users current accomplish tasks without the benefit of the proposed product. Ideally, the users surveyed should have some experience in using their own digital data at the national level. Subjects addressed should include:

1. What line of financial reasoning led to using digital national data?
2. How are digital data now accessed, without a national database?
3. Which factors of success can enhance financial benefits?
4. Which data afford the greatest benefits?
5. How will user environments change as a consequence of the creation of a national database?
6. What do users consider to be the ideal future situation?

These six subjects and similar matters are conveniently handled in interviews, using a standard form.

Procesing Data acauired in the first User Study

Business criteria should be used in analyzing the material compiled in the first user study. That is, cost-benefit analyses are conducted using net present values of incomes and costs, which in turn involves choices of a discount rate and a time horizon. The discount rate should be the same as that of projects of corresponding risk. The time horizon customarily is ten years.

11.5.3   DEFINING OFFERINGS

Knowledge of the behaviour of today's market permits estimation of tomorrow's activities. In other words, products may be specified in terms of data content, accuracy, completeness, currentness, distribution, costs, operation, financing, pricing, etc.

A structured attack is used to specify products suited to identified market needs. The aspects of various products may be conveniently ordered in a matrix. Each item may be entered in three categories of ascending service, as geographic accuracies of ± 2 m, ± 10 m, ± 50 m, or updating every three months, once a year, or once every five years, etc. A typical matrix, for a national road database, is shown in the table of Fig. 11.18.

In theory, the eight rows and three columns of the matrix of Fig. 11.18 permit over 6,000 different varieties of products. However, few users will be able to select among and rank more than 30 varieties. Consequently, the varieties should be culled down to a selected few.

11.5.4   COST MODEL

A cost-benefit analysis presupposes that a cost is associated with each item in a product matrix. A realistic cost model should relate to a basic product. The model may then be used to compute the cost of the basic alternative of the

	LOW SERVICE	MEDIUM SERVICE	HIGH SERVICE
A. GEOMETRY	Inter-regional All public roads No private roads	Intra-regional All public roads All forest roads	All-to-all All public roads All private road
B. ATTRIBUTES	Road identity Road class Road number Public ferries	Cruising advice Speed limits Critical slopes Weight/height etc.	Advanced advice Addresses Traffic lights Zip code etc.
C. GEOMETRIC ACCURACY	Low precision +/- 50 m x,y	Medium precision +/- 10 m x,y	High precision +/- 2 m x,y,z
D. CURRENTNESS	Revised every 5 years	Yearly Revised annually	Continuous 1 - 3 months
E. CORRECTNESS * OF ATTRIBUTES	>95% of all data	>99% within 1 yr	>99% right away
F. USER ACCESS	Magnetic medium national format	On-line access without GIS-supp.	Full GIS support on-line access
G. DELIVERY TIME TABLE	Normal cycle within 8 years	Speeded within 3 years	Crash program within 1 year
H. PRICE	Low	Medium	High

*) Public road network data correct by 99,9% including attribute data road identity and restriction, within time limit 1 - 3 months.

Fig. 11.18 - Typical product matrix, for a national road database.

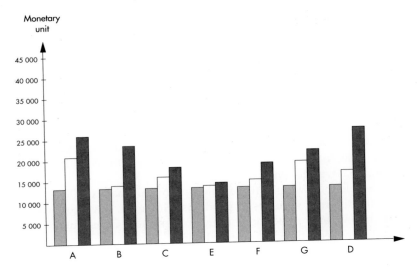

Fig. 11.19 – Effects on cost by partially increasing attribute support levels. for A – G ref. Fig. 11.18.

product and to compute the marginal costs of the other alternatives. A typical example of this procedure is illustrated in Fig. 11.19.

11.5.5   MARKET ANALYSIS: SECOND USER STUDY

The goal of the second user study is to measure the reactions of potential users to the various products proposed. The users involved in the first user study should interviewed again. Conjoint analyses are employed in conducting the interviews and processing the data acquired.

Conjoint analyses involve the users ranking products on the basis of simultaneous evaluations of all properties of each alternative. In practice, the product alternatives are stated on "cards," which users first divide into four groups:

1. Very good
2. Relatively good
3. Relatively poor
4. Very poor

In the next round, users rank the "cards" of each group on a scale of 1 to 9. This then permits calculating the relative importance of the items in the product matrix. A typical example is shown in the table of Fig. 11.20.

The next step is to produce curves for each item (A through H in Fig. 11.20) that illustrate the profiles of choice of each user group. A typical example is shown in Fig. 11.21. Note that the inclination to pay for products is vital in a marketing profile.

Item	User group: 1	2	3	4	5	6	Average	Rank
A: Network content	9	23	14	18	14	26	17	2
B: Attribute content	9	13	9	10	10	9	10	6
C: Geometric accuracy	6	15	10	9	17	22	13	4
D: Currentness	35	11	22	24	12	14	20	1
E: Error-free	10	5	11	9	7	12	9	7
F: Data accessibility	7	7	6	8	9	9	8	8
G: Completion date	20	12	14	10	5	3	11	5
H: Price	6	16	13	13	28	9	14	3

Fig. 11.20 - Relative importance, example of a Norwegian national road database.

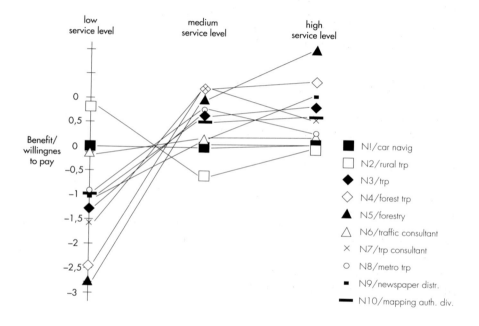

Fig. 11.21 - Example of market profile curves.

Whenever the expressed needs of the users are spread on the ordinate of the plot and the profile lines cross, then goals coincide. An unvarying curve indicates an item of little interest. A steep curve indicates an important item for which a user is willing to pay the price of going from a lower to a higher level of service. These market profiles are used to match products to the market.

The interviews of selected, experienced users may be supplemented with broader surveys of potential users conducted by mailing out questionnaires.

11.5.6   DELINEATING STRATEGIES

Strategies must suit prevailing market conditions. But as market conditions change, prudent strategies must encompass alternative approaches with respect to:

Organization on the production side, as:
- decentralised approach where users initiate data acquisition to meet their own needs
- centralised approach in which the organisation further processes its own data by incorporating new data
- semi-decentralised approach in which the organisation is responsible for reasonable standardised data interchange between providers and users

Market situation, as:

- individual approaches, none joint, and a varying, heterogeneous user group
- joint market activities due to common interests in limited local geographic areas
- joint market activities, cost sharing for map data and centralised attribute data
- major user market demand for new GIS products, but inadequate for fully financing costs
- large, homogeneous market demand; near monopolisation of data provision possible

The question of public contra private involvement also enters the evaluation picture, as do questions of regulation and accessibility. Economies of scale and market homogeneity also influence choices of strategies.

Public agencies, which often operate national databases, may choose among various business approaches, as:

1. Supplement basic data with other standard data and distribute the combination as a standard package to all users (extended map production).
2. Supplement basic data with other standard data to compile tailor-made user products.
3. Supplement basic data with user-specified data to compile tailor-made user products.
4. Coordinate data interchange without being involved in producing, but with standardisation responsibility.

Product price should depend on risk. Joint market ventures can lower prices.

Financing through national budget appropriations may increase the availability of products to users, but may also hold prices artificially low so providers cannot meet costs. If private financing is to be involved, the market must tolerate competition and price variation, which may exclude some users who cannot afford the products offered.

Socioeconomic considerations dictate that public databases should be accessible to all, with few limitations.

11.5.7 EXAMPLE OF A STRATEGY FOR A NATIONAL MAP SERVICE

Recommended strategies for national road databases in Scandinavia have:

General strategy

The national map services should adopt an offensive strategy to adapt to rapidly changing technologies and markets.

Creating databases

National road databases created with data from various sources should be aimed towards a small nucleus group of users willing to pay for the services provided. This will make the project viable. This product cannot classify as a public service as many financially weaker groups will be excluded.

Currentness

Currentness is paramount, that is, updating every three months for important items and at least once a year for all other items.

Financing

A multi-theme road database should be user financed. As the market demand is heterogeneous, the product is not a characteristic public service. Semi-public financing may gradually increase during the updating phase, as products become standardised and more users access the data.

The portion of the road database dedicated to rationalisation of national map production can be financed by normal public funds.

Pricing

Market pricing should involve a high entry price in order to cover updating costs. Otherwise, the road data will be inadequately financed, which will create problems in the updating phase and pricing problems for the initial group and for subsequent users.

In the long run, the most sought data will undoubtedly become a public responsibility, a fact that may influence marginal costing in pricing.

Organisation

The national map services operate both as administrators of public map policies and as producers of products for the commercial market. Consequently, the map services should be divided into two units. The basic, public service unit will administer the geometrical information (infrastructure). A more commercial unit will compete in the private market, and should focus on value adding activities.

Access to basic data

All private and public organisations should have equal access to basic data that is produced in connection with national map production. In this way, provider and user interests balance, and various interests may compete on equal footing.

12   GIS IN DEVELOPING COUNTRIES

12.1   PRESENT SITUATION AND THE NEED FOR GIS

Information is one of the most important strategic factors influencing development. Modern political and economic systems cannot function without a continuous interchange of reliable information. Worldwide, the information sector is large and rapidly growing. Geographic information plays an increasingly important role, both nationally and internationally; it is decisive in the environmental monitoring and management essential to the survival of future generations [see Section 14.6 of Chapter 14 on the GRID system].

In many developing countries, efforts thus far has focused on campaigns to satisfy short-term acute needs. Investments that have yielded immediate tangible results, as in food production, infrastructures and health, have taken precedence over other investments, which have been downgraded or neglected. The regrettable result in most developing countries is that overviews of natural resources and environmental status are meagre. The information requisite to planning and prudent management of renewable and other resources is inadequate. There are no reliable overviews of the extent of environmental damage caused by human activities.

According to The World Commission on Environment and Development, as reported in Our Common Future [Oxford and New York, Oxford University Press, 1987, paperback 400 p., ISBN 0-19-282080-X] sustainable development is the key to economic growth in development countries. Sustainable development is defined [Our Common Future, p. 46] as "a process of change in which the exploitation of resources, the direction of investments, the orientation of technological development, and institutional change are all in harmony and enhance both current and future potential to meet human needs and aspirations."

One of the barriers to sustainable development is that most developing countries lack the information requisite to planning it. Typically, the depletion of natural resources accelerates, which causes deforestation, overgrazing, soil erosion and extinction of an unknown number of animal and plant species (reduction of the genetic multifariousness). There is not enough information to permit computing the effects of climatic changes which may be triggered by global warming. Nonetheless, GIS models permit rudimentary computations. The effects can be dramatic, as shown in Fig. 12.2 which show the effects of water level increase in the Nile Delta.

Uncontrolled urbanisation is a growing problem in many developing countries. Lack of information is one of the greater barriers to dealing with the problem. Characteristic for the situation is that epidemics, wars and other

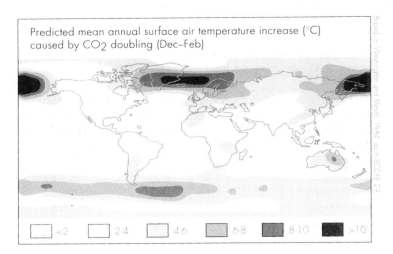

Fig. 12.1 - Predicted temperature increase

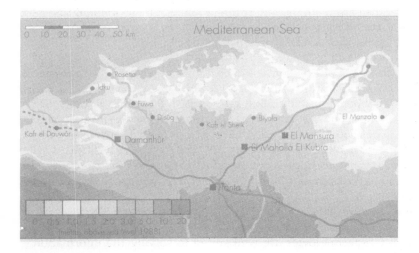

Fig. 12.2 - Effect of Nile delta water level increase.

factors have concentrated populations in areas with few available resources. Infrastructures, as buried cables and pipes, are poorly documented in cities, which complicates repairs and, in turn, impairs electricity and water supply.

The lack of information in developing countries may be stated in financial terms. As of the early 1990s, the total geographic information sector accounted for approximately 0.5% of the GNP in most industrialised countries. For most African countries, the corresponding figure was 0.1% of far lower GNPs. The resultant paucity of information hinders political and economic development.

History underscores the connection between information and development. Many African countries once were well mapped, as the colonial powers sought to administer and exploit the natural resources of their colonies. In most cases, this information resource has not been maintained.

In many cases, data are compiled for project purposes with no regard to other uses. The qualities of these data collections and the acquisition methods employed are frequently unknown. The result is that data available are unwieldy and difficult to collocate in productive ways.

Recent UN agency and World Bank studies have underscored the need for improved geographical information in developing countries worldwide, particularly in Africa. The potential gain may be illustrated by comparison. Fully 92% of Europe is mapped in a scale of 1:25,000, which permits detailing of environmental themes. Only 2% of Africa is mapped in a scale of 1:25,000 (F. Falloux, 1992).

Thus far, central decision makers have not emphasised the need for environmental data. This lack of political interest coupled with inadequate overviews of natural resources result in uninformed decision making. Financial aid organisations have thus far been only moderately helpful, as their involvements have usually been limited to short-term projects of limited geographic extent.

Although the need for information is universally acute, most developing countries lack the financial resources and professional capabilities to acquire and manipulate data. The geographic information technologies may help resolve the dilemma. The use of satellite data provides a rapid and cost-effective means of covering larger areas. Entering data in GIS entails structured, standardised storage which simplifies retrieval and collocation. GIS tools may be used to execute both simple and complex tasks, and data may be presented on map sheets or in reports. GIS is expected to play a key role in the development of geographic information in developing countries in the 1990s.

Even though GIS technologies are well suited to solving the information problems of developing countries, the high-tech approach may involve problems due to:

- political circumstances
- cultural traditions
- management conditions
- lack of expertise
- inadequate infrastructure
- extreme climates
- insufficient financial resources

12.2   STRATEGIES FOR INITIATING GIS IN DEVELOPING COUNTRIES

12.2.1   RANKING

Initiating GIS is not itself a goal, but rather a means to attain other basic goals. Consequently, superjacent principles and goals should guide the application of GIS and other high-tech facilities in developing countries. Sustainable development could be one such superjacent goal. Another might be to mitigate the plight of the poor through land reform. In all cases, a developing country should have a clear policy for, or at least a position on the use of GIS.

In Chapter 11, inception studies were identified as essential to the successful implementation of GIS in organisations. This holds true for extensive implementations of GIS facilities in countries. Likewise, to assure cost-effective, user-oriented systems, benefits must be assessed in terms of the goals delineated through user studies involving decision makers, planners, managers and others and their needs for information.

In developing countries, the primary users of geographic information frequently are unfamiliar with GIS and its benefits. Therefore, GIS implementations should start with pilot projects that aim both to provide experience and to rapidly meet short-term, high-benefit information needs.

Without an implementation strategy, projects in developing countries often become technology-driven, and technology takes the upper hand, ahead of the tasks it is to perform. The result may be an unprofitable overinvestment in hardware and software at the expense of primary tasks, which remain undone.

Numerous studies have shown that prudent GIS implementation can be cost-effective in developing countries. A good example is that of a study recently conducted in Lesotho.

Lesotho is a southern African kingdom completely bounded by South Africa, with an area of 30,460 km² (about that of Belgium), an approximately square geography - 200 km from North to South and 230 km from East to West, and a population of 1.48 million (1984 estimate). The economy is based on intensive agriculture and contract labour by men working in South Africa. Much of the land suffers soil erosion, and the country remains dependent on food imports. Diamonds are the only significant developed natural resource, although there is prospecting for coal, oil and uranium.

A study of the need for introducing GIS at the national level in Lesotho examined 80 development aid projects of various types. All 80 project were found to depend on geographic information to varying extents. Fully 50 projects produced geographic information in the form of thematic maps, statistics related to geography, and so on. The projects had acquired information through detective work in various institutions in the country and abroad, often at considerable expense in time and money. The information generated by a project was usually filed in reports which were essentially inaccessible to others. In some cases, the reports were discarded upon completion of the project.

Clearly, if projects were required to store geographic information in standardised form, the benefits accrued through rationalisation could be enormous. The data could then be used to rapidly build national geographic information databases.

In Lesotho, and in other developing countries, geographic data are often associated with:

- general map making
- environmental data centres
- natural resource monitoring
- network mapping
- urban development

A prudent first step to instituting GIS at the national level in a developing country is to contact UN agencies and programmes, as the United Nations Environmental Programme (UNEP), the Global Resource Information Database (GRID) and United Nations Institute for Training and Research (UNITAR), as well as the World Bank, for advice and strategy notes and compendia of similar projects in many countries.

12.2.2   CHOICE OF TECHNOLOGY

The technology and the technology transfer involved in a GIS implementations should satisfy defined goals and suit user needs and local conditions. In addition, the technology employed in a GIS facility should be integrated in a long-term program of resource management, in order to avoid ad-hoc solutions of questionable longevity.

The choices of hardware and software should be limited to recognised products of known quality, to ease further development and compatibility and to simplify national and international cooperation. This obvious principle has sometimes been ignored, particularly in bilateral projects in which donor countries have specifically favoured their own GIS products, despite their being second-rank.

Technology transfer should be progressive. That is, competence should be increased before more advanced technology is brought in.

Comprehensive testing should be conducted to assure that the individual units (computers, plotters, digitisers etc.) function individually and in concert in the system for which they are intended.

In general, the equipment chosen should be robust, of recognised production and maintainable locally. Long-term maintenance agreements are preferable for the sake of system reliability. Local agents usually are preferable for speed of finding faults and providing service. In many cases, a single supplier with total equipment responsibility may provide the least expensive and most reliable service.

Ambient conditions, as unstable mains power supply, climatic extremes (heat, humidity, dust), must also be considered in planning.

12.2.3   BUILDING CAPABILITIES

Effective education is universally essential. If a country is to successfully exploit GIS technology, its officials and planners must fully fathom the implicit political, social and economic potentials. Likewise, there will needs for research workers to develop and adapt technologies, engineers to develop applications systems, technicians to operate systems and instructors to teach the uses of the technologies. The needs for all these skills necessitates a series of training programmes, at various levels and with differing goals and specialisations.

The International Institute for Aerospace Survey and Earth Sciences formally the International Training Centre (ITC) of The Netherlands, which is experienced in training programmes in developing countries, recommends that for GIS technologies to succeed, minimum personnel training should comprise the following groups (J.L. van Genderen, 1991):

- Decision makers and planners, including officials and administrators, who need a general understanding of the practical possibilities and limitations of GIS as a decision making tool.
- Leading personnel in institutions, public management agencies and private companies, who need sufficient technical knowledge to coordinate the introduction of GIS.
- Personnel involved in the practical surveying and acquisition of digital data used in GIS.
- Technicians responsible for the operation and maintenance of equipment and programs.
- Research workers knowledgeable in GIS and with expertise in applications development and in GIS as an analytical tool.
- Instructors responsible for training and teaching the various categories of personnel and knowledgeable in GIS technologies and their practical applications.
- School pupils and university students.

As a rule-of-thumb, when GIS is introduced in an established organisation, a tenth of the staff should go through primary training. When new organisations are established, about 30% of the staff should be trained in GIS per year.

Training may be conducted either by sending personnel to an industrialised country or by sending experts to a developing country. Training in an industrialised country is often expensive and therefore limited in the number of personnel it can accommodate. However, it is often the most effective, as relevant equipment and well-qualified instructors are often available.

Training may be arranged in the form of courses at various centres, as in-house courses or as on-the-jobtraining. Courses for skilled personnel should last two to four weeks. Basic theoretical and practical courses in GIS should last six to twelve months. Further training then can be in GIS-related topics at universities in industrialised countries.

For decision-makers, shorter in-house courses and workshops of five to six days duration can be effective.

Training in the uses of GIS in natural resource management is now offered by UN agencies and programmes, including the Global Environment Monitoring System (GEMS), United Nations Institute for Training and Research (UNITAR), the United Nations Environment Programme (UNEP), the Global Resource Information Database (GRID) and the Food and Agricultural Organisation of the United Nations (FAO). The International Institute for Aerospace Survey and Earth Sciences formally the International Training Centre (ITC) in The Netherlands and many universities also offer courses, as do regional and national training centres, as Asian Institute of Technology (AIT) in Bangkok.

12.2.4 ORGANISING AND COORDINATING

In addition to functioning technically, a GIS facility should contribute to national development. Consequently, the introduction of GIS in a developing country must be seen in the national perspective, with a view towards a rational organisation and towards coordinating the activities of dissimilar sectors. Most developing countries must start from scratch, which can actually ease the evolution of a sensible organisational structure. Isolated sectorisation and monopolisation of information should be avoided.

Some sectors of GIS activities, as mapmaking, may advantageously be organised jointly by public agencies and private firms, so as to avoid building an inefficient monopoly.

Providers may have idealistic, governmental or partly business goals. User organisations are oriented towards development, education, work places and political conditions, and in many instances represent special interests at the national level.

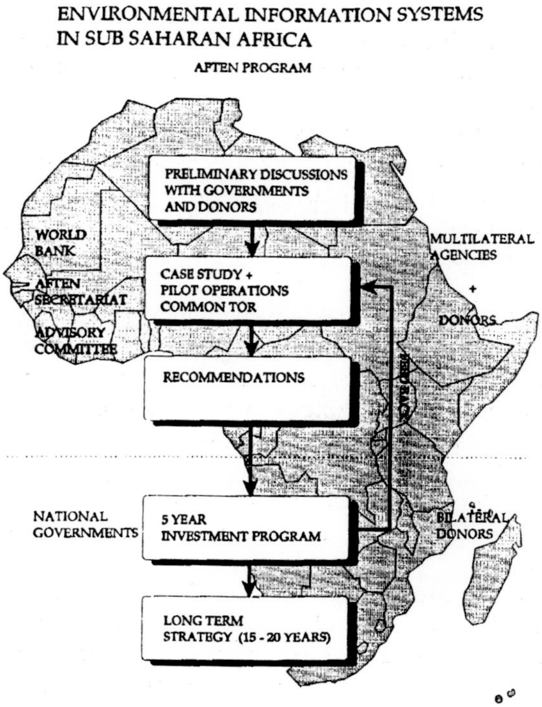

Fig. 12.3 - NEIC Uganda.

Technology transfer usually should be based on the cooperative work of several organisations, as bilateral and multilateral aid organisations, product suppliers, schools and universities, map making organisations, consultants, and so on.

These factors indicate that the chief goal of a donor organisation should be to support the primary needs and goals defined in a recipient country.

The introduction of GIS usually entails multi-sector activities, as cooperation between utilities (water, sewage, electricity, telecommunications) or directorates (roads) in matters pertaining to planning, development and management of infrastructures. A current example is in Uganda, which is now establishing a national environmental information centre based on GIS technologies. The Department of the Environment was responsible for the inception study which identified seven Ministries as producers and users of environmental information.

In developing countries, the challenges involved in introducing GIS often are more institutional and organisational than purely technical. Orchestrating cooperation between providers and users, so the GIS facility fulfils its purpose, may be more demanding than commissioning the system itself.

A successful introduction of GIS requires the cooperation of GIS experts and sector experts, a facet which should be reflected in the institutional organisation of GIS.

The current trend is towards recommending that GIS centres for processing environmental data be organised as independent entities under the leadership of boards or managerial groups in which the major users are represented. Such a centre can function only if it is supported by a network of actively supportive institutions.

At the national level, binding governmental support is requisite to the long-term commitment involved in building geographic information databases.

Projects often must be organised to suit the particular management practices in a developing country. Political instability and frequent reorganisation of public services can complicate the introduction of GIS.

In the early 1990s, the World Bank initiated a program to develop Environment Information Systems (EIS) in all of Africa south of the Sahara. The participating countries have a common approach to problems:

1. Identification of users of environmental data and ranking of needs.
2. Identification of institutional, financial and legal impediments.
3. Development of a long-term strategy for establishing an environmental information system, including a detailed plan for the first four years.
4. Realise the plan.

Uganda is a participating country, in which the work is organised almost like a private venture and is lead by a local expert, with support provided through periodic visits by experts from an industrialised country (Norway). All labour-intensive tasks, as acquisition of data in the field, are conducted in Uganda by local personnel. Several pilot projects have been conducted, dedicated to energy, national parks and cadastres. Training and briefing of governmental decision makers is an important part of the overall activities.

12.2.5   FINANCING GIS IN DEVELOPING COUNTRIES

As discussed above, developing countries spend little on the acquisition and processing of geographic data, only about 0.1% of their GNPs. One reason for the low level is that there are no markets capable of sustaining the costs. In the face of acute, short-term needs, governments have not been able to allocate sufficient funds for building up information products, as maps and databases. The result is a vicious circle, as information is requisite to solving most acute problems that the countries repetitively face.

As is the case in industrialised countries, the procurement of geographic information in developing countries is basically a public sector task, the costs of which cannot be borne by the initial users. The problems of imperfect markets can be resolved only if governments and aid organisations appreciate the necessity of securing the necessary basic information.

The first phase of such an effort is for poorer countries to request full financing by aid organisations. Use fees may be introduced gradually to offset costs, but for the first ten years they cannot be expected to offset more than 10% to 25% of the aggregate costs of initial installations, operation and maintenance of a facility. However, if a system isstructured in response to needs identified in user surveys, the socioeconomic benefits may far exceed investments.

The World Bank Technical Department Environment Division Africa Region (AFTEN) has established a secretariat dedicated to introducing GIS and remote sensing for improving resource and environmental management in some forty countries south of the Sahara. This work is coordinated with a corresponding programme initiated by UNEP and GRID. The time horizon of the programme is 20 years, and it is divided into five-year periods for progressively building local capabilities and for introducing technologies.

In short, implementing GIS in a developing country involves far more than a single isolated project of limited duration. Implementing GIS must be based on suitable long-term strategies, without which the experience of many unsuccessful projects has proven that it may fail.

13   ELECTRONIC CHART DISPLAY AND INFORMATION SYSTEM (ECDIS)

13.1   INTRODUCTION

Dependable marine navigation has always relied on good nautical charts. International agreements, as those under the authority of the International Marine Organisation (IMO) and/or the Safety Of Life At Sea (SOLAS) convention, delineate a myriad of requirements germaine to safety. Among the more important requirements are stipulations that relevant, current nautical charts, lighthouse lists, tide tables, and the like shall be readily accessible on board. Nonetheless, increasingly severe accidents occur. As ship sizes are ever increasing, the consequences of a single accident can be catastrophic, as when the Exxon Valdez supertanker ran aground on 24 March 1989 in the archipelago south of the Valdez oil terminal on the Gulf of Alaska east of Anchorage. Close to a quarter million barrels of raw oil spilled into the sea, polluting it and the countless miles of shoreline throughout the Gulf, causing extensive damage to wildlife and the fragile ecology at 61°N. Thus far, the cleanup has cost two billion US dollars. More common than such environmental disasters is the frequent loss of life in marine accidents. From a purely technical viewpoint, the litany of marine accidents is borne out by world accident statistics: incorrect course and/or speed cause over half of all accidents. The risk of marine accidents is apparently larger than the risks accepted in other comparable sectors, as in aviation.

Part of the problem outwardly is that despite the widespread use of radar and other navigational aids, the quality of the navigational information available on board most ships has not kept pace with ship sizes, speed, complexities and values.

Traditionally, ship positions are plotted on nautical charts. So traditional plots are continuous records of previous positions. The process has gradually become semi-automatic, using electronic aids to navigation: positions are now plotted using data from navigational aids, as position receivers and radars on the bridge. The maps used are usually updated or supplanted, using written reports and new maps supplied by various national chart agencies.

Fully automatic systems that resemble marine mobile GIS are now available. They are known as   Electronic Chart Display and Information Systems (ECDIS).

13.2   ECDIS BASICS

An Electronic Chart Display and Information System incorporates three information sub-systems:

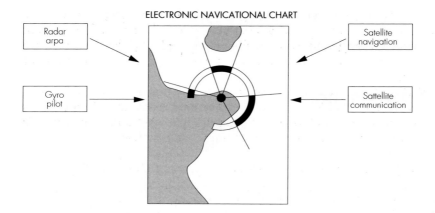

Fig. 13.1 – Electronic navigation is based upon the integration into one system of electronic navigation chart, global positioning system, satellite communication, radar and gyro pilot.

An Electronic Navigation Chart (ENC) which combines nautical information, as lighthouses and markers,and hydrographic information, as coastlines and sea bottom topography, for display on screen. The basic information is usually digitised from traditional nautical charts. Position information from the ship's radar and positioning system. Steering information from the ship's steering system.

ECDIS provides a complete electronic navigation and warning system that displays the ship position and other information vital in navigation. An ECDIS is usually implemented in a PC.

An ECDIS can automatically and continuously, day and night:

- Display a symbol of the ship in real-time motion relative to navigational obstructions and other traffic in surrounding waters.
- Display the symbol of the ship in the same scale as the features of the surrounding waters; i.e., show a real image of the geometries involved.
- Select and display selected map details of interest with respect to weather, surrounding waters and the size and operational characteristics of the ship.

Provide audible and visible warning signals in the event that:

- the course of a sailing or a planned sailing is charted through non-navigable waters (anti- aground)
- ship deviates from planned course
- a planned change of course is not made
- ship does not deviate from a course of potential collision with other moving objects (anti-collision)

If radar images are entered into the ECDIS, other vessels and their manoeuvring in nearby waters may be displayed. If an autopilot is connected, the ECDIS can automatically follow a planned course. Changes of course are then made after the navigator enters an acknowledgement of a warning of course change.

An ECDIS on board reduces the manual operations involved in manoeuvring in heavily trafficked waters, and consequently affords the ship's officers more time for monitoring ship progress and other traffic. In turn, this reduces the risk of running aground or colliding with another vessel.

Users may implement several GIS functions in an ECDIS, including:

- increase displayed map scale in arduous waters
- divide the screen display into windows, as to simultaneously display neighbouring map sheets
- display only selected information on screen
- perform computations, as sailing distances
- enter own geometry and attribute data

As is the case in GIS, an ECDIS user can build up a database containing descriptions (attributes) of objects shown on digital maps. The descriptions entered include lighthouse lists, sea bottom conditions, fish catches and the like. New navigational and hydrographic information may be entered on the digital maps. The new information may comprise customary updating, or may be specific to an application, as a fisherman entering the positions of equipment set out and the locations of wrecks that affect catches.

Global Positioning System (GPS) and other positioning system equipment may be integrated with an ECDIS. As discussed in Section 6.6.1 of Chapter 6,

DIFFERENTIAL – GPS

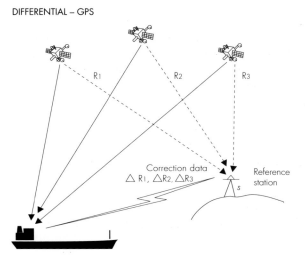

Fig. 13.2 – Use of a differential GPS in electronic navigation.

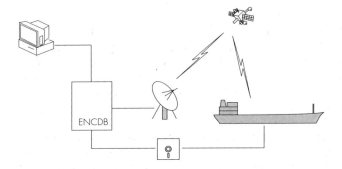

Fig. 13.3 – Updating of an Electronic Navigation Chart Data Base (ENCDB) is based on satellite communication and data input from diskettes.

the normal accuracy of GPS is ± 100 m. Accuracies better than ± 10 m, as may be needed in narrow channels, are available in differential GPS, which requires coast reference stations. The accuracies provided by GPS and differential GPS are better than those of many older nautical charts. This implies that accurate nautical charts are a prerequisite to the sensible use of GPS.

Updating is essential to the successful use of ECDIS, as navigational conditions continually change. National cartographic and hydrographic agencies regularly issue updating information, which may be entered into an ECDIS. Tests are now (early 1992) being conducted on transmitting updating information to ships via the INMARSAT satellite communications system.

In the late 1980s (1989 -90), the Norwegian Hydrographic Service conducted a comprehensive three- year ECDIS project. Titled SEATRANS, it included trials of fully-integrated ECDIS on the bridges of ships in regular traffic between Trondheim in Norway, Hamburg in Germany and Amsterdam in The Netherlands. These trials corroborated the utility of ECDIS, and showed that its most useful functions are:

- display of route sailed
- planning of coastal and channel sailing
- following planned route
- display of clear waters
- selection of various map details, as lighthouses and their sectors
- verification of position

13.3   ELECTRONIC NAUTICAL CHARTS

Digital navigational and hydrographic data are compiled by vectorising data from scanned nautical charts, by manual digitising of nautical charts and by hydrographic surveys.

Area symbols are used on conventional nautical charts. Therefore, digital nautical charts must include area information that can be readily identified, as being coloured in a display, to support the ECDIS anti- aground function. The goal is for the ECDIS to trigger an alarm should a ship approach or plan to approach a dangerous area, as one too shallow for it to pass safely. The quality of the depth data can be enhanced considerably if all areas around reefs, skerries, shallows and major wrecks are marked and identified with area symbols entered in ECDIS.

Nautical charts must be of extremely high quality, as users must be able to assume that the data they contain are geometrically accurate and as free of error as possible. Consequently, official nautical chart data must thoroughly verified before being issued.

As is the case for conventional topographic maps, transferring conventional nautical chart data to digital form entails numerous problems. Spaghetti data must be corrected to delete loose ends and close gaps. All polygons must be closed and assigned. As indicated in Fig. 13.4, the correction tasks can be considerable when a map grid square may contain upwards of five to six thousand polygons.

Reading conventional nautical charts is a skill which must be learnt. The skilled chart reader interprets details and intuitively associates objects related to each other, as depth figures for shallows, texts pertaining to islands and reef numbers designating reefs. Data processing systems do not possess such skills. Consequently, interpretations and intuitive connections must literally be spelled out for an ECDIS, in terms of computations, through topology or in stated attributes.

To a great degree, basic nautical chart data are oriented in chart sheets. Adjacent sheets customarily overlap, to ease reading from sheet to sheet. When chart sheet data are converted to digital form, the overlapping areas of adjacent sheets may give rise to double entries, which complicate the admini-

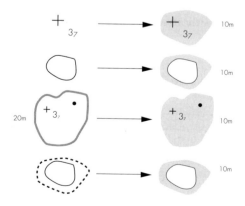

Fig. 13.4 – The relationships between the objects and area shading are important elements in an Electronic Navigation Chart.

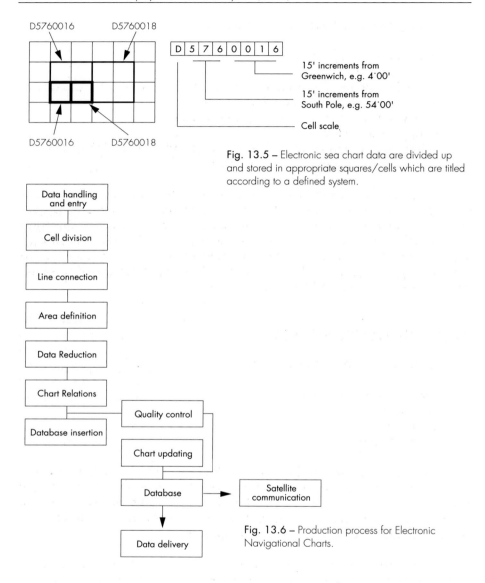

Fig. 13.5 – Electronic sea chart data are divided up and stored in appropriate squares/cells which are titled according to a defined system.

Fig. 13.6 – Production process for Electronic Navigational Charts.

stration of databases. Therefore, prior to conversion to digitalform, data are divided into cells varying in size from 7.5 minutes square to 8 degrees square, depending on the original chart scale and the amount of data contained within each cell. In digital form, a single cell can contain up to 7 Mbytes of data. Retrieving cells containing such large volumes of data may be a slow process. However, trials have shown that the volume of data in a cell may be shrunk by up to 95% using box- spline mathematics, with no sacrifice in quality.

The data should be divided into various themes:

- information always on screen
  - ship draught
  - coastline, underwater reefs and skerries, bridges, suspended cables, etc.
- information which also should be standard
  - ferry routes, sailing markers, restricted areas, etc.
  - supplementary information
  - detailed descriptions of dangerous points and areas, updating date, compass declination and magnetic dip

An international data format, the International Hydrographic Organisation (IHO) DX90, has been developed for the interchange and updating of electronic navigational information. The IHO object catalogue contains a thorough structuring of data and definitions of all objects of interest in maritime applications. The catalogue also contains guidelines on how objects are to be digitised. As an example of the extent of detail, an object designated reef with peak marker is defined by 34 different attributes.

13.4   OUTLOOK FOR ECDIS

As of the early 1990s, ECDIS is used on board:

- ferries, coastal vessels, etc.
- fishing vessels
- pleasure craft
- military vessels

Systems range in complexity and price. From raster based systems, slightly more than an inexpensive PC upwards. The cheapest systems are too simple and not approved by the International Maritime Organisation or the International Hydrographic Organisation, and consequently are recommended only as aids to other navigational equipment.

The potential market is considerable. On the basis of current (1992) registries and newbuildings of ships over 1,000 GRT, the annual ECDIS market should reach 500 units by 1995 and more than 1800 units by 1999. The improved navigation provided by ECDIS is essential to some shipping sectors, as the development of large, high-speed passenger vessels.

As this book goes to press (early 1992), ECDIS has been used as an aid to navigation in relatively limited channels and coastal waters and at sea, where high navigation accuracy is not necessary. ECDIS is so new a technology that statutes relevant to it have not yet been worked out, and the various national hydrographic surveys have yet to acquire the equipment needed to produce the myriad of complex information required.

ECDIS can be truly global only if national hydrographic services are capable of producing authorised information covering the most important, most trafficked and most dangerous channels and coastal waters. Worldwide, an estimated 2,000 nautical charts must be converted to digital form to cover the most important ports and sailing lanes.

Plans are now being drafted to establish an international electronic chart centre in Norway, which will support databases capable of supplying all ECDIS users with basic information and updatings.

14. GIS FACILITIES: EIGHT EXAMPLES

14.1 OVERVIEW

Since the late 1970s, GIS facilities of various sizes, complexities and purposes have been implemented and have been proven in service. This chapter comprises summary descriptions of eight facilities, selected to illustrate the applications and scope of GIS. As this book goes to press (early 1992), these facilities have been in operation from 4 to 15 years, are operated by both public and private organisations, and range from dedicated project facilities to global resource facilities. They are:

Cities: multi-user facilities, mostly for municipal services:
- Oslo, Norway
- Edmonton, Alberta, Canada
- Burnaby, British Columbia, Canada

Utilities: facilities serving specific systems:
- Detroit Edison (electricity supply), Detroit, Michigan, USA

Multidisciplinary Engineering: dedicated facilities for major projects:
- Central Railway Station, Stockholm, Sweden

National and Regional Planning Agencies: resource facilities for planning services:
- National Physical Planning Agency, The Netherlands
- Geodatasenteret A.S., Norway

Global resource facilities:
- Global Resource Information Database (GRID), under the auspices of the United Nations

14.2 CITIES

The density of georeferenced information in cities is far greater than that in towns or rural areas. Consequently, city agencies and bureaus were among the first to implement GIS facilities, primarily to support municipal services. And as many of them often apply the same information, GIS facilities in cities typically feature collective data and promote inter-agency coordination.

14.2.1 OSLO, NORWAY

Oslo is Norway's capital, located at 60° N at the northern end of the Oslo Fjord. With a population of 452,000 in the city itself (one-tenth of the populationof the country) and 650,000 the metropolitan area that includes the adjoining suburbs, Oslo ranks small compared to cities elsewhere. But the city proper covers an area of 454 km², half again as large as London and more than four times as large as Paris. The topographical features of the city reflect those of the country: Oslo has forests, mountains, lakes, rivers and farmlands. Some 27 km² of its area comprises the tip of the Oslo Fjord and its numerous inlets. In addition to the customary functions of the seat of government and centre of finance of a country, Oslo has a harbour and industries. Small but highly visible portions of the city's activities are in farming, forestry, fishing, hunting and outdoor recreation. Oslo measures some 40 km from its northern to its southern border, and 20 km from east to west. Only a quarter of the total area is fully urbanised. Understandably, Oslo has needs for GIS that are greater than its population count otherwise indicates.

Traditional georeferenced municipal information in Oslo has long been plagued by problems typical for cities:

- Information is spread among agencies and stored in a variety of dissimilar ways, which escalates the time involved in access and retrieval.
- Duplication, both in data acquisition and storage, is widespread, which is costly.
- Currentness is poor, as updating is spotty, which requires that excessive amounts of data be compiled and verified for each use.

These difficulties led Oslo to initiate a GIS pilot project in 1984. The pilot project lasted two and a half years and resulted in a seven-year inception project to implement a GIS facility. The Town Planning Office, Computer Services Bureau, Soil Mechanics Division Office, Electricity Board, Surveying Services, Water Supply and Sewage Works, and Public Roads Administration participated in the projects.

The strategy employed emphasised collective data of common interest to all municipal organisations. The collective data were defined as comprising:

- base map database for technical applications (scales of 1:500 to 1:10,000)
- base map database for overviews (scales of 1:10,000 to 1:25,000)
- routings of piping and cables with type codes, for use in water supply and sewage, electricity supply and road works
- subsurface maps with bedrock contours and drilling depths
- development maps showing type and geographical coverage
- register associated with development maps

These collective data are increasingly used in:

- planning and processing of construction permits
- management of cable and pipe networks
- area resource management
- map production
- planning and designing
- property information
- works coordination
- road and traffic information

Data specific to individual agencies, as operational information and the details of networks, are not of common interest and therefore are stored and maintained by the agencies.

The pilot project concluded with an extensive evaluation of benefits. With a time horizon of 25 years and a discount rate of 7%, the overall benefit-to-cost ratio was found to be 1:1, exclusive of the considerable benefits to the public and external users. The benefits for the various user groups are shown in Fig. 14.1.

The concept of collective data and inter-agency coordination proved formidable in practice. The professional paradigms and imbued routines of the various agencies could not be changed sufficiently rapidly to fully accommodate the new concepts. Hence the original strategy of a collective GIS was modified to a distributed approach in which each agency created its own database with all users having access to all databases.

No one system was found which could fully satisfy the diverse needs of the various agencies. Consequently, the initial strategy was changed to permit the individual agencies to select their own systems after consultation with the core project staff. The result has been that four or five different systems are now in use, including the Norwegian SysScan and the American ArcInfo. However, data are organised to permit recombination. Hence, if and when a collective system is implemented, data defined as collective may be transferred directly from the agency databases to a collective database.

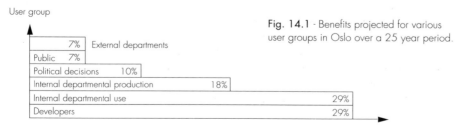

Fig. 14.1 - Benefits projected for various user groups in Oslo over a 25 year period.

Existing base maps were scanned to rapidly provide a body of digital map data. Compressed raster data was found to require no more storage capacity than vector data. The long-range goal is to produce thematic vector data for the entire city.

14.2.2   EDMONTON, CANADA

Edmonton, the capital of the Province of Alberta in Canada, is a city of 564,000 about 280 km northwest of Calgary. Until 1947, the city was primarily an administrative, educational and agricultural centre, aswell as a node for railway lines to the vast farming regions of the Province. In 1947, oil was found nearby in Redwater, Pembina and Leduc, and oil now dominates the city's economy, with refineries, petrochemical industries and raw and finished plastics producers. The municipality is enormous, covering an area of some 1800 km², only part of which is urbanised. Edmonton is nearly self-sustaining, with its own water and electricity supplies and sewage and refuse processing works.

Following two years of planning, a GIS pilot project was initiated in 1977. The immediate goal was to switch from traditional means to GIS to cut map production costs through increased productivity, as well as to provide decision support in area planning, demographic analyses and works planning. In the first seven years, compilation focused on data concerning:

- properties
- roads
- addresses
- piping and cable networks
- land use
- property assessment

The pilot phase concluded in 1978. Thereafter a distributed database system was built up, with individual databases in the various organisations, interconnected by a network to provide mutual access.

In 1984 the project was thoroughly reviewed and evaluated; three main conclusions resulted:

- A full-time project manager must be hired.
- The participating agencies must be reorganised.
- GIS should be developed in stages emphasising coordination and collective use of data.

When complete, the property database, which is the system's map base, will cover both the city of Edmonton and the entire province of Alberta in 1500 maps of scale 1:1,000.

The Office of the City Assessor employs data on taxable objects (properties and buildings) that are used in collecting taxes and in budgeting. The system ensures a more rapid, more accurate assessment of taxation than previously possible.

The Planning and Building Department uses the system to update municipal addresses, for drawing and area computations in zoning plans and to produce various thematic maps (population data, building ages, land uses, etc.). The zoning plans include overviews of properties affected, which are used in giving notice of imminent works. The Agency also maintains 1:5,000 scale maps and produces various specialised graphic products.

In 1981 Edmonton Power installed GIS facilities which are now used to mapping suspended and buried cables, street illumination, channels and overhead trolley wires for trolley buses. These overviews are used for planning and other purposes, and more than 400 maps and reports are generated every year.

The system aids operation and maintenance by producing network overviews for inspection and maintenance crews, schematic overviews of splices, and special overviews for hospitals, shopping centres, schools and universities. The system is also used to produce standard construction manuals including complete lists of materials and to compile overviews of subscribers.

The Transportation Department has used GIS mainly to compile maps (property boundaries to roads) and for entering details on roads and streets, as traffic signals, traffic islands, kerbs, etc., as well as easements (use rights). Road and street planning maps are produced on the basis of these data. The system is also used internally to produce specific products as floor and office plans, organisation plans and various other graphic products.

In 1980, Edmonton Telephone started its own pilot project, which concluded that GIS was useful, but fully so only after the entire telecommunications region was covered. Information on cables from existing map sheets and technical and fiscal data for all exchanges have been entered. Work orders and other information are used to continuously update the data, as with overviews of cable lengths and current status of conduits and poles.

The Water and Sanitation department uses the system mostly in connection with the operation and maintenance of the water supply and sewerage networks. Other uses include producing detailed construction plans, overviews of services for users, analyses of network faults (1,000 to 1,500 per year), as well as various graphic products as organisation plans.

The overall benefit-to-cost ratios of the implemented GIS facilities in Edmonton are from 2.5:1 to 4.0:1. After the first ten years of operation, it became clear that system operation and maintenance and other costs could be cut by replacing the older, original complex and expensive work stations with newer, simpler models.

The experience gained in Edmonton has shown that automatic map production is more than twice as fast as manual map production, and updating of information is from five to 20 times faster. In addition, new products have been made available and information is both more accurate and more readily available.

Converting and entering data to create databases was originally planned for and thereafter executed over a period of many years. However, dealing with both conventional and digital data over a longer time period proved cumbersome. Therefore, basic map database compilation was given priority.

The distributed database approach, with databases in the various organisations has proven to be awkward. Although all the databases may be accessed via a network, there is no concise overview of the data available in all the organisations. Consequently, several types of data, as data in the Transport Ministry's road database, will be made collective, a step which will increase the uses of and benefits from data.

The basic system program was compiled primarily for mapmaking purposes. Consequently it did not support operations involved in manipulating polygons, computing distances, overlaying and other more advanced tasks. More "intelligent" programs are installed, and communications between the various databases and computers are being upgraded.

Historical data, which are needed in many planning tasks, should be maintained in databases.

14.2.3   BURNABY, CANADA

Burnaby is a city in the Province of British Columbia on the west coast of Canada. It is located on the mainland, to the east of and contiguous with Vancouver, and is bordered on the north by the Burrard Inlet on the south by the Fraser River, the major river of mountainous British Columbia. Burnaby has a population of about 140,000 and covers an area of about 500 km². The Simon Fraser University was founded there in 1963, and the city is a communications hub, with the Trans- Canada Highway and the Canadian Pacific Railway. Municipal services, as police, fire brigade, health services, parks and planning are the responsibility of The Corporation of Burnaby.

In 1977, The Corporation of Burnaby started a GIS pilot project jointly with municipal engineering departments (electricity, gas and telephone). The pilot project goal was to find how the participating organisations would benefit from an automatic map system. The outcome was favourable, so in 1978 a project was initiated to create a database.

The project started with a Synercom ST 100 software system run on a PDP 11/750 computer. In 1984 the system was upgraded with three separate VAX 11/750 computers and a VAX 7500 CPU interconnected by an Ethernet. The goal of the system was to support automatic map production, maintain map data, create an attribute database and produce technical drawings.

As illustrated in Fig. 11.14, the project organisation was large and effective.

Previous experience showed that even simple transactions often involved several agencies and that the personnel involved often needed to have access to the same basic information. A detailed study showed that nearly 70% of the daily civic work load involved data on properties: georeferencing and attribute data were needed in all works pertaining to roads and streets, water supply, sewage, buildings, parks and the like.

All the agencies were involved in the inception phase of the project and all agreed on the strategic choices made.

The digital databases contain data on property boundaries, zoning plans, zoning ordinances, assessment facts, addresses, elevation contours, buildings and roads and streets. Data on the water supply network and sewerage are entered. The databases also contain information on parks and greenery, and some criminology statistics have been entered from The Canadian Bureau of Census.

In addition to being employed in management, daily work and planning, the data have been used in new ways, including locating a new fire station and in studying population density and age distributions in parts of the township.

For the first three years of the project, total costs were found to exceed benefits by one million dollars. However, in the following four years benefits increased so greatly that after the project had been in operation for seven years, net benefits were close to half a million dollars per year.

The efficiency attained can be illustrated by example. In the pre-GIS days, base maps were updated manually, at a labour cost of approximately one man-year and with a consistent backlog of six to twelve months. Using the GIS facilities, the labour cost has been cut to 1.5 man-months and the backlog reduced to two to four months. One of the ancillary benefits to the city has been that income from property taxes has increased due to the upgraded overviews of properties.

The various combinations of digital map data and attribute data have resulted in new applications, including:

- optimum route descriptions for fire brigade calls, based on information on the road system, traffic volume and speed limitations
- analyses of traffic volume and traffic controls as used in transportation planning
- planning of parks and recreational areas on the basis of population data and information on existing park areas
- planning of refuse transport

In addition to the municipal agencies originally involved in the project, the Building Inspection Department, the Financial Office and the Cultural and Recreation Department also used the system.

14.3   ELECTRICITY SUPPLY: DETROIT EDISON, USA

Electric power companies usually plan, install and maintain the electric lines and cables and other facilities needed to distribute electric energy over geographic areas. So many power companies are experienced in automatic line and cable routing and have recently implemented GIS facilities for planning and operating their networks.

In general, electric power companies employ dedicated GIS facilities with little coordination with other organisations. There are two reasons for the trend. One, electric power companies were among the first to employ GIS, and consequently implemented GIS for their own uses. Two, mandates, agreements or other political factors may hinder collaboration with others.

Detroit Edison is a public company responsible for the transmission and distribution of electric energy in the southeastern part of the state of Michigan in the USA. The population of the 20,000 km² area served is 4.9 million, and Detroit Edison has 1.7 million subscribers and 11,000 employees.

The prime goal of the GIS project was to transfer 20 to 30 dissimilar line and network map series to a common digital information system for circuits, to streamline planning, operation and maintenance. In 1975 the project began with a pilot phase. By the end of 1980, all maps for four of the six areas covered were entered in the system. Thereafter, complete databases for these four areas have been compiled, and data for the remaining two areas were entered. Detroit Edison is responsible for the operation and maintenance of the system.

The system supports some 30 products and services, many of which were not envisioned when the GIS project started. The more important products and services include:

- Distribution circuit maps (transmission lines, distribution grid)
- Freeway construction maps
- Lighting maps
- Incandescent light maps
- Airport pole height maps
- Pole audits
- Engineering statistical package
- Facility maps
- Industrial sites marketing maps
- Municipality index maps

- Plant accounting reports
- Pole test maps
- Sewer system maps
- 10-year distribution master plan maps

The system had overcome the backlog on updating of 2,000 man-years and cut updating time to two to three weeks. Otherwise, the time savings were dramatic. Some products, which previously required 75 hours of work, could be produced by the system in a minute and a half.

Dismissals due to the rationalisation afforded by GIS have been few. The man-hours saved by GIS have been transferred to project-related activities.

At the end of 1980s he entire GIS project had cost more than $ 30 million. The resultant annual savings amount to 100 man-years, of which 35 are in circuit mapping. In addition, the company pays less tax on its equipment and facilities, as the system provides more accurate and more current overviews.

Circuit details and other terrain information are entered with an accuracy of ± 1 m in densely built-up areas and with an accuracy of ± 3 m in rural areas. Detroit Edison is responsible for acquiring base maps and employed orthophotography in many areas.

During the course of the project, the need for geometric accuracy in design was found to be greater than originally assumed.

GIS data acquisition, storage and presentation are mainly for internal needs, as Detroit Edison has scant cooperation with municipal organisations and other utilities, as the telephone company, in acquiring base information. Consequently, cooperation and collective data may be cost-savers of enormous potential.

The early emphasis during system implementation was on structuring data and on workable databases, a historical fact that the project managers hold responsible for the success attained and for the many new applications that have been generated. Another contributory factor is that the project staff has been stable and dedicated to the task.

There have been some encouraging peripheral, non- technical results. In many ways, the company found that it is easier to train staff to use the GIS circuit information system than the previous manual circuit information maps. Staff turnover has gone down as the GIS facility has come on-line.

14.4 ENGINEERING APPLICATIONS: STOCKHOLM CENTRAL STATION

GIS facilities are employed in a wide range of projects, as in residential developments, road building, airport planning and the like. However, GIS has typically been used not in an entire project, but only in parts of it, as overview planning, optimising bills of quantities, visualising, and so on. Consequently, many of the potential benefits of GIS have seldom been realised in projects.

The ways that a more extensive use of GIS can benefit a project are well illustrated by the Stockholm, Sweden Central Railway Station Extension Project.

Stockholm, the capital and largest city in Sweden, lies on the Baltic coast at the outflow of Lake Mälar. The city covers an area of 186 km² and has a population of 651,000; the population of the greater Stockholm metropolitan area is 1.6 million. In addition to the many governmental institutions, Stockholm is Sweden's financial centre, is the see of a bishop, and has many educational and cultural institutions, including the Nobel Institute. It is also an industrial centre, particularly in engineering services, electronics and metal working. Railways play an important role in the communications of the city and the adjoining districts. Stockholm is both a hub and a terminus on the 11,200 kilometre Swedish Railway system, which is among the more extensive of the OECD countries: annual usage corresponds to 716 passenger-kilometres per head, 40percent more than in Great Britain and nine times that of the United States or Canada. The Stockholm Central Railway Station was first built in 1871 and subsequently rebuilt and expanded several times, most extensively in 1984 through 1988, to increase platform capacities, incorporate a new main hall for local and intercity traffic, and add new mail and bus terminals. The GIS facility used a system delivered by NordCad AB of Sweden, run on a McDonnel-Douglas GDS system.

The reconstruction and expansion project involved many disciplines which needed information on existing facilities, as tracks and switches, towers and poles, platforms, manholes, overhead conductors, buried cables and conduits, and so on. All the relevant data were entered in a central map database, which then supported design. Various consultants also used the system to document designs.

The project was divided into nine rebuilding stages, each with a "rebuilding map," which delineated the works to be done. All computations requiring georeferenced parameters were executed rapidly, as all the information needed was available from the central database. Maps were also continuously produced to meet the needs of various design tasks.

In all, the maps and other documents generated comprised some 60 standard maps for each of the nine rebuilding stages, 300 special-purpose maps and more than 3,000 impromptu reports containing georeferenced information.

The use of computerised georeferenced information in a project of this magnitude was found to offer numerous advantages, including:

- The project stage maps that illustrate progress can be both more detailed and geometrically more correct, which eases further design.
- Errors and omissions were more rapidly located, as all information was available a central database.
- Interactive design of railway tracks, towers and platforms, with access to all georeferenced information, saved considerable time compared to conventional design.
- Appreciable time was also saved in all computations of coordinates, distances and other parameters, as well as in superimpositions and drawings.
- All completed works were measured, and the measurements were entered in the system, which permitted rapid verification against plans. When the rebuilding and expansion project was finished, a complete description of the station area was available in the database.

Otherwise, the project, with its many close deadlines, illustrated the benefits of concise data structuring and precise dataflow.

14.5   NATIONAL AND REGIONAL PLANNING

National planning agencies have traditionally processed statistical data at the national level. A larger portion of these data are georeferenced. Consequently, many national planning agencies have incorporated GIS for processing and presenting these data. The National Physical Planning Agency of The Netherlands, is typical.

14.5.1   NATIONAL PHYSICAL PLANNING AGENCY OF THE NETHERLANDS

Rijks Planlogische Dienst, The National Physical Planning Agency is located in the city of Zwolle (pop. 88,000), the capital of Overijssel Province in the eastern part of The Netherlands. Since the early 1970s, the Agency has been responsible for the operation and maintenance of national geographic databases, which are available for national and regional planning.

The system uses a central Prime 9955 computer located in Den Haag, 150 km from Zwolle. Various software systems are used, including the American ArcInfo.

Three databases have been created:

- Township database: data based on maps in scale of 1:250,000 and other sources.
- Grid (raster) database: data with cell sizes varying from 100 X 100 m to 2 X 2 km.

- Topographic database: base maps in scale of 1:25,000.

The Township database contains 3,500 different sets of data for the 800 townships in The Netherlands, as population data, building data, agricultural data and environmental data. Historical data are included, which permits studies of particular years or analyses over several years. Data for the townships may be combined to support regional analyses.

The main themes contained in the database are:

- population, by sex, age and civil status
- births, by sex, status and mother's age
- removals to/from, by sex, age, civil status and pattern
- travel patterns
- building data, by age and type
- land use
- farm production
- labour market, by occupation

The results of analyses using these data are usually presented on 1:250,000 scale overview maps that cover the entire country, such that townships and regions may be compared.

The Grid (raster) database contains mostly data concerning climate, soils, geology, landscape features and pollution.

A sub-database created previously contains all addresses in The Netherlands. Each address is linked to type of house and year of building and is located on the national grid, in a 100 X 100 m cell in densely built-up areas or a 500 X 500 m cell in less densely built-up areas. This permits computing building density, building age and population density for each cell, throughout the country. This information is used in localisation, environmental, supply and other matters. The grid is also used in ecological studies of birds, animals, trees and plants.

The Topographic database contains all the information on maps of scale 1:25,000, as well as attribute data, as numbers of beds in hospitals, traffic data, number of boats on canals and lakes, biomass production in shallow water areas, and so on. Full topology is used in the database, so the data it contains may be used in a variety of analyses.

The database has already supported numerous analyses, including:

- traffic noise studies
- charting the transportation of gas, petrol, explosives and other hazardous materials

- locating power lines and windmills
- developing recreational areas

The quality of planning in The Netherlands has increased as a consequence of using GIS at the regional and the national level. This will eventually benefit the citizens of the country.

14.5.2   GEODATASENTERET A.S., NORWAY

In the late 1980s, regional service centres for geographic data were established in several countries. For instance, by 1991 there were four such centres in Sweden and three in Norway. The purpose of a centre usually is to provide regional expertise in the production, distribution and use of geographic information. Geodatasenteret A.S. (Geo-data Centre Ltd.) was the first such centre in Norway.

Geodatasenteret was established in the third quarter of 1987 in Arendal, Norway, as a joint venture between governmental agencies, including the national map service, and various private firms. The centre has no staff of its own but it is runned by hired staff. All project activities are conducted by the owner organisations. However, the centre is responsible for purchasing and maintaining specialised equipment for GIS and remote sensing. It derives its income from hiring this equipment to the owner firms, whose individual equipment budgets are inadequate for purchase of expensive, specialised items. The centre also provides geographic data to the region, as well as providing equipment services that enable firms to conduct projects that otherwise not possible.

In 1991, the staff immediately available to Geodatasenteret numbered 35 professionals, and the centre performed several major national and international projects.

The Swedish Teragon system is used for digital image processing, and the American ArcInfo system is used in GIS applications. A Calcomp colour raster plotter is used to plot images and maps.

In the GIS sphere, the centre has concentrated on developing applications for land use planning and environmental data. Detailed plan data, as road geometries, are retrieved from CAD systems and digital terrain models for further GIS uses. Overview data are retrieved from national and regional databases. Digital planning data are used dynamically in the planning processes of GIS analyses and for producing several special products, as statistical data for planning areas, thematic maps and the like. Production routines have been developed for overview plans and detail plans, and include digital reproduction using Postscript, Scitex and other systems.

The centre has created a collective information system for six townships in the Arendal region. The information system supplies geographic data for business development and tourism, as well as routine geographic information. The information entered include data on infrastructures, business properties, schools, kindergartens, tourist destinations, etc. The system may also support area planning and environmental analyses. An ancillary dedicated system is used for population statistics.

The centre's remote-sensing activities concentrate on vegetation mapping using processed images from the SPOT and LANDSAT satellites and from scanned aerial photos. These activities have included extensive studies of forest vitality. Colour enhancement, supervised and unsupervised classification and segmenting techniques are used in image analyses. The centre performs extensive mapping of developing countries, using satellite images.

Image processing and GIS have been connected, to permit the use of map information in image processing and vice versa.

The successful operation of the centre has shown that cooperation between governmental agencies and private firms is beneficial. The physical proximity of staff members and the joint use of specialised equipment have created a qualified professional culture. In turn, this has attracted new firms and ventures, including the regional United Nations office for the GRID coverage of Scandinavia and the Pole areas.

14.6   GLOBAL RESOURCE INFORMATION DATABASE (GRID)

The global environmental problems are undeniably large and complex. Up until the mid 1980s, most environmental problems were addressed individually and locally. However, as the underlying problems are global, there clearly is a need for the global approach now possible using computer technologies. Consequently, in 1985 the United Nations Environmental Programme (UNEP) established the Global Resource Information Database (GRID) as part of an overall effort to collect, manipulate and promulgate global environ-

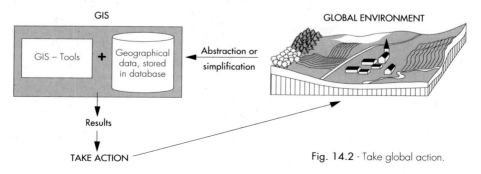

Fig. 14.2 - Take global action.

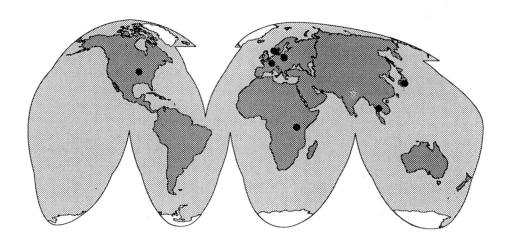

Fig. 14.3 – GRID Network

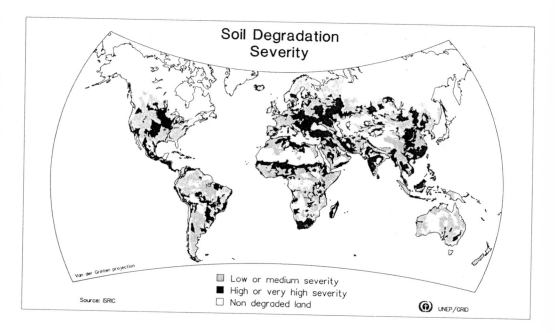

Fig. 14.4 - Soil degradation severity (GRID).

85° N, 80° E

66° N, 0° E

Fig. 14.5 – Ice distribution Barents sea. Extremes for February from 1966-1988. (GRID-Arendal)

Shift of Boreal Zone in Europe with 2 x CO2
OSU Scenario (Biome 1.1)
Potential for boreal / nemoral vegetation

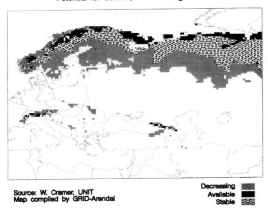

Source: W. Cramer, UNIT
Map compiled by GRID-Arendal

Decreasing
Available
Stable

Fig. 14.6 – CO_2 affect on vegetation.

mental and natural resource data. By using GIS, remote sensing, telecommunications and other modern technologies, these data may be made readily available to professionals, government officials, decision makers, managers and the general public. GRID's mandate is to acquire data and information relevant to prudent management of the global environment. GRID is also charged with training of developing country personnel in GIS and remote sensing.

The United Nations Environmental Programme (UNEP) is responsible for coordinating the environmental work of all United Nations organisations and for furthering environmental activities at all levels. Part of UNEP's work is conducted through the Global Environmental Monitoring System (GEMS), which has its headquarters in Nairobi and is responsible for coordinating and adapting global environmental data. GRID is part of that network.

When fully implemented, GRID will be a global network with centres in all hemispheres and with national service centres in many countries. By the third quarter of 1991, seven GRID centres were established: Nairobi in Kenya, Geneva in Switzerland, Bangkok in Thailand, Arendal in Norway, Warsaw in Poland, Tsukuba in Japan and Sioux Falls, South Dakota in the USA. Many countries are involved in financially supporting GRID, and computer hardware and software suppliers have donated considerable amounts to the programme.

GRID itself compiles little basic data, but relies mostly on data from cooperating institutions (UN organisations, other international organisations, national organisations, etc.). These data are evaluated and processed by scientists of various disciplines and then aggregated and sent to GRID. Data concerning climate, biodiversity, fresh water, coastal pollution, renewable resources and environmental health protection are of special interest in GRID work. The data compiled are analysed and structured in GIS facilities for graphical presentation and other purposes. This process involves evaluations of how point observations are best presented in areas on maps. By 1991, GRID had created over 30 sets of digital global data, including political boundaries, natural boundaries, topography, soil types, vegetation, hydrology, precipitation, temperature, ozone distribution, population density and endangered species. In addition GRID has compiled individual national datasets and data sets for Africa. GRID data are tax free and free of charge to all users.

GRID is also building up a reference database which contains information on content, owner, accessibility and other details of all existing databases of interest to GRID work.

GRID conducts individual GIS and remote sensing projects. One of the strengths of GRID is that it may be used to study the correlations between differing sets of data. Studies have included assessments elephant populations in Africa, modelling of vector-borne diseases in Africa, studying the effects of climatic changes on sea level and plant production, evaluating and mapping forest resources in Europe and in Ghana and creating an environmental database in Uganda. GRID has also played a central role in the compilation of a global atlas of desert spreading, using data from about 400 diverse sources.

The GRID centre in Arendal, Norway has developed a number of specialised products, including

- an environmental and natural resource information and presentation system
- a reference database for the Arctic
- an animated, dynamic presentation of the ice cover in the Barents Sea for the past 20 years, to facilitate long-term observations of interest in environmental evaluations

- presentations of sulphur and pH in air, soil and waters in Norway
- long-distance transport of air pollutants in Europe
- scenarios of affects of CO^2 on vegetation and soils

GIS is the most important tool for analysing and presenting data as those compiled by GRID. In the future, multimedia techniques, as animation, will undoubtedly be employed to make information more readily available to the citizens of the world.

REFERENCES

CHAPTER 1

Aronoff, S. (1989). "Geographic Information Systems: A Management Perspective". ISBN 0-9218404-00-8, WDL Publications, P.O. Box 585, Station B, Ottawa, Ontario K1P 5P7, Canada.

Burrough, P.A. (1986), "Principals of Geographical Information Systems for Land Resource Assessment", Oxford Science Publications, ISBN.

Cowen, David (1987), "GIS vs. CAD vs. DBMS "What are the differences?", Proceedings GIS '87 - San Francisco.

Crosswell, Peter L. and Clarc, Stephen R., "Trends in automated Mapping and Geographic Information System Hardware", PE and RS, No. 11, 1988.

Dale, P. F. & McLaughlin, J. D. (1988). "Land Information Management. An introduction with special reference to cadastral problems in Third World countries", ISBN 0-19-858404-9, Clardon Press, Walton street, Oxford GB.

Dangermond, Jack (1988), "GIS Trends", Arc News, ESRI, California.
Dangermond, Jack, Monrehouse, Scott (1988), "Trends in Hardware for Geographic Information Systems", Environmental Systems Research Institute.

Goodchild, M. F. & Kemp, K. K. (1990). "National Centre for Geographic Information and Analysis. Core Curriculum: Introduction to GIS. Technical issues in GIS. Application Issues in GIS." University of California, Santa Barbara, USA.

Hamilton, Angus og McLaughlin, John (1986), "Land Information management education", FIG-congress 1987, Toronto.

Maguire, D. J. & Goodchild, M. F. & Rhind, D. W. (1991). "Geographic Information Systems". ISBN 0-582-05661-6, Lougman Scientific & Technical, Burat Hill, Harlow, Essex CM20 2JE, England.

Marble, D.F. Calkins, H. Peuguet, D.J. Brassel, K. og Wasilenke, M. (1981), "Computer software for spatial data handling (3 Vols)". Prepared by the International Geographical Union, Commision on Geographical Data Sensing for the US Departement of the Interior, Geological Survey. IGU, Ottawa, Canada.

Morgan, John (1987), "Academic Geographic Information Systems Education: A commentary". PE and RS no. 19, 1987.

Nordic KVANTIF (1987). "Community Benefit of Digital Spatial Information. Final report". Viak, Arendal, Norway.

Spandau, L. and Kasperidus, H (1988), "Research of the ecological system of Brechtesgaden - main influence on ecological systems in the high mountain area of the alpine national park of Brechtesgaden". ESRI-User conference, München 1988.

Strand, Geir Harald (1988) "Adoption of GIS in Europe", Norsk Regnesentral.

Tomlinson, R.F. (1984), "Geographical information systems – a new frontier". In Proc. Int. Symp. on Spatial Data Handling. 20-24 August, Zürich pp. 1-14.

Tomlinson, R.F., Calcins, H.W. og Marble, D.F. (1976), "Computer handling of geographical data", UNESCO, Geneve.

CHAPTER 2

Bernhardsen, Tor (1984), "Samfunnsmessig nytteverdi av kart og geodata-GAB (KVANTIF)". Rapport fra NTNF-prosjekt. Rapport nr. V, VIAK AS.

Bernhardsen, Tor (1985), "Samfunnsmessig nytteverdi av kart og geodata – det tekniske kartverket i kommunene (KVANTIF)". Rapport fra NTNF-prosjekt. Rapport nr. IV, VIAK AS.

Eliasson, Anders (1987), "Digital Kartbearbetning". Kompendium til påbegynnerkurs i fjär-analys". Sveriges Lantbruksuniversitet, Umeå.

Goodchild, M. F. & Kemp, K. K. (1990). "National Centre for Geographic Information and Analysis. Core Curriculum: Introduction to GIS. Technical issues in GIS. Application Issues in GIS." University of California, Santa Barbara, USA.

Zuart, P.R. (1986), "The production of information for policy decisions from land information systems", FIG-congress 18, 1986, Toronto.

CHAPTER 3

Burrough, P.A. (1986), "Principals of Geographical Information Systems for Land Resource Assessment", Oxford University Press, New York.

Cederholm, T. & Persson, C-G. (1989). "Standardiseringsverksamheten i Sverige inom GIS-området - terminologifrågor, informasjonstrukturering och kvalitetssekring". ULI, Gevle, Sweden.

Goodchild, M. F. & Kemp, K. K. (1990). "National Centre for Geographic Information and Analysis. Core Curriculum: Introduction to GIS. Technical issues in GIS. Application Issues in GIS." University of California, Santa Barbara, USA.

CHAPTER 4

Aronoff, S. (1989). "Geographic Information Systems: A Management Perspective". ISBN 0-9218404-00-8, WDL Publications, P.O. Box 585, Station B, Ottawa, Ontario K1P 5P7, Canada.

Bedard, Yvan (1986), "A study of the nature of data using a communication based on conceptual framework of land information systems", FIG-congress 1986, Toronto.

Burrough, P.A. (1986), "Principals of Geographical Information Systems for Land Resource Assessment", Oxford Science Publications.

Carter, James R., "Digital Representations of Topographic Surfaces", PE and RS, no. 11, 1988.

Cook, B.G. (1983), "An introduction to the design of geographic databases". In Proc. Workshop on databases in the natural sciences, CSIRO Division of Computing Research 7.-9. September 1983, pp. 175-186, Cunningham Laboratory, Brisbane, Queensland.

Dale, Petre, McLaughlin, John (1988), "Land Information Management", Claredon Press, Oxford.

DeSimone, M. (1986), "Automatic structuring and feature recognition for large scale digital mapping", Auto Carto London, Vol. 1, pp 86-95, London.

Goodchild, M. F. & Kemp, K. K. (1990). "National Centre for Geographic Information and Analysis. Core Curriculum: Introduction to GIS. Technical issues in GIS. Application Issues in GIS." University of California, Santa Barbara, USA.

Maguire, D. J. & Goodchild, M. F. & Rhind, D. W. (1991). "Geographic Information Systems". ISBN 0-582-05661-6, Lougman Scientific & Technical, Burat Hill, Harlow, Essex CM20 2JE, England.

Robb, Margareth (1986), "Raster Geographic Information Systems, State of the Art", Notat, Norsk Regnesentral.

Strand, Geir Harald (1986), "The map as a Spatial Data Base", Weltan and environment, Tapir Forlag, Trondheim.

CHAPTER 5

Aronoff, S. (1989). "Geographic Information Systems: A Management Perspective". ISBN 0-9218404-00-8, WDL Publications, P.O. Box 585, Station B, Ottawa, Ontario K1P 5P7, Canada.

Bedard, Yvan (1986), "A study of the nature of data using a communication based on conceptual framework of land information systems", FIG-congress 1986, Toronto.

Burrough, P.A. (1986), "Principals of Geographical Information Systems for Land Resource Assessment", Oxford University Press, New York.

Dangermond, Jack, Monrehouse, Scott (1988), "Trends in Hardware for Geographic Information Systems", Environmental Systems Research Institute.

Goodchild, M. F. & Kemp, K. K. (1990). "National Centre for Geographic Information and Analysis. Core Curriculum: Introduction to GIS. Technical issues in GIS. Application Issues in GIS." University of California, Santa Barbara, USA.

Goodchild, Michael (1988), "A spatial analytical perspective on geographical information systems", Int.J. Geographical Systems nr. 4, 1987.

Hageny, Wolgang (1988), "Computergraphic".

Maguire, D. J. & Goodchild, M. F. & Rhind, D. W. (1991). "Geographic Information Systems". ISBN 0-582-05661-6, Lougman Scientific & Technical, Burat Hill, Harlow, Essex CM20 2JE, England.

CHAPTER 6

AML (1987), "Alternative modeller for ledningsregistrering", del-rapport 1, 2, 3, 4, HLO-hovedstadsregionens Ledningsoppmåling I/S, København.

Andersen, Ø. (1980), "Fotogrammetri".

Andersen, Øystein (1980), "Fotogrammetri grunnkurs-bind 2", Institutt for landmåling, Norges Landbrukshøyskole, Ås.

Aronoff, S. (1989). Geographic Information Systems: A Management Perspective". ISBN 0-9218404-00-8, WDL Publications, P.O. Box 585, Station B, Ottawa, Ontario K1P 5P7, Canada.

Becker, J-M. (1988). "Treghetsposisjonering". Lantmeteriverket, Sweden.

Bolstad, P. V. & Gessler, P. & Lillesand, T. M. (1990). "Positional uncertainty in manual digitized map data". Int. Journal of Geographical Information Systems, no. 4.

Brugard, Bengt (1981), "Kart på Data", Universitetsforlaget.

Bråten, K. (1986). "Fjernanalyse", Viak, Norway.

Burrough, P.A. (1986), "Principals of Geographical Information Systems for Land Resource Assessment", Oxford Science Publications.

Carter, J. R. (1988). "Digital Representation of Topographic Surfaces". Photogrammetric Engineering & Remote Sensing, no. 11.

Chrisman, Nicholas (1987), "Efficient digitizing through the combination of appropriate hardware and software for error detection and editing". Int. J. Geographical Information Systems, nr. 3, 1987.

Cinon, N. & Coe, P. K. & Quigley T. (1990). "A regression technique for estimating the time required to digitize maps manually". Int. Journal of Geographical Information Systems, no. 4

Dale, P. F. & McLaughlin, J. D. (1988). "Land Information Management. An introduction with special reference to cadastral problems in Third World countries", ISBN 0-19-858404-9, Clardon Press, Walton street, Oxford GB.

Dæhlen, M. & Holm, P. G. (1991). "Datakompresjon". Ingeniørnytt nr. 44.

Eliasson, Anders (1987), "Digital Kartbearbetning-kompendium til påbegynnerkurs i fjär-analys". Avdeling for fjär-analys, Sveriges Lantbruksuniversitet, Umeå.

ESRI (1990). "Understanding GIS. The Arc/Info Method". ESRI, Redlands California, USA.

Forsberg, R. (1988). "Inertial surveying methods". Geodetisk Institute, Denmark.

Goodchild, M.F. & Kemp, K. K. (1990). "National Centre for Geographic Information and Analysis. Core Curriculum: Introduction to GIS. Technical issues in GIS. Application Issues in GIS." University of California, Santa Barbara, USA.

Hultquist, Nancy & Scripter, Sam (1987), "A new GIS - but what about "old" data". Proceedings, GIS '87 - San Francisco.

Hunter, G. (1985), "Archival data and land information systems". University of Melbourne, Dept. of Surveying technical report, Parkville, Australia.

Larsson, Ingemann, Revan, Peter (1986), "Datamaskinen og samfunnet", NKI-forlaget.

Maffini, G. (1990). "The role of Public Domain Databases in the Growth and Development of GIS". Mapping Awareness, no. 1.

Ortaawa, Tom (1987), "Accuracy of digitizing: Overlooked factor in GIS operations", GIS '87. San Francisco.

Oliver, M. A. 6 Webster R. (1990). "Kriging: a method of interpolation for geographical information systems" Int. Journal of Geographical Information Systems", no. 4.

Rainio, A. (1990), "The geo-data dictionary". FIG XIX, comm. 3, Helsinki Finland.

Sanktjohanser, Gabriell (1987), "ERDAS – ArcInfo Interface Applications", ESRI User Conference München, 1987.

Smith, S. M. & Scheider, H. & Wiart, R. (1987). "Agricultural field management with micro-computer based GIS and image analysis systems". GIS`87 vol. I, San Francisco USA.

Strande, Kari (1989), "NGIS-nasjonalt geografisk informasjonssenter", Kart og plan nr. 2-1989.

Taxt, T. (1991), "Digital gjenkjenning av tekst og geometri", Kartdagene 1991, Norway.

Thompson, C.N. (1984), "Test of digitizing methods", OEEPE Official Publication no. 14, IFAG, Frankfurt.

Van Roessel, J.W. og Fosnight, E.A. (1984), "A relation approach to vector data structure convension", In Proc. Int. Symp. on Spation Data Handling, 20.-24. august, Zürich, pp. 78-95.

Warner, W. S. & Carson, W. W. & Braaten O. M. (1991). "GIS data capture from aerial photographs: A case study". Photogrammetric Record, April 1991.

Wolf, Paul R. (1983), "Elements of Photogrammetry", McGrew-Hill.

Østensen, O. (1991), "Hva skjer på standardiseringsområdet najonalt og internasjonalt. Hva gjør Kartverket". Kartdagene 1991, Norway.

CHAPTER 7

Andersen, Ø. (1991). "Geografisk Informasjons-System. Innføring. Presisering. Problemstilling". Norges landbrukshøgskole, Norway.

Aronoff, S. (1989). "Geographic Information Systems: A Management Perspective". ISBN 0-9218404-00-8, WDL Publications, P.O. Box 585, Station B, Ottawa, Ontario K1P 5P7, Canada.

Dunn, R. & Harrison, A. R. & Wite, J. C. (1990). "Positional accuracy and measurement error in digital databases of land use: an empirical study". Int. Journal of Geographical Information Systems, no.4.

Goodchild, M. F. & Kemp, K. K. (1990). "National Centre for Geographic Information and Analysis. Core Curriculum: Introduction to GIS. Technical issues in GIS. Application Issues in GIS." University of California, Santa Barbara, USA.

Heuvelink, G. B. & Burrough, P. A. & Stein, A. (1990). "Propagandation of errors in spatial modelling with GIS". Int. Journal of Geographical Information Systems, no. 4.

Maguire, D. J. & Goodchild, M. F. & Rhind, D. W. (1991). "Geographic Information Systems". ISBN 0-582-05661-6, Lougman Scientific & Technical, Burat Hill, Harlow, Essex CM20 2JE, England.

Welch, Stephen m.fl. (1987), "Recognition and Assessment of Error in Geographic Information System", PE and RS, no. 10, 1987.

CHAPTER 8

Aas, A. G. & Mathiassen, H. J. (1989). "Aktuelle geografiske informasjonssystemer". FGIS-prosjektet, Statens Kartverk, Norway.

Abel, D.J. (1983), "Towards a relational database for geographic information systems". Workshop on Database in the Natural Sciences, CSIRO Division of Computing Research 7-9. September 1983. Cuningham Laboratory, Brisbane, Queensland, Australia.

Andersen, Ø. (1991). "Geografisk Informasjons-System. Innføring. Presisering. Problemstilling". Norges landbrukshøgskole, Norway.

Aronoff, S. (1989). Geographic Information Systems: A management Perspective". ISBN 0-9218404-00-8, WDL Publications, P.O. Box 585, Station B, Ottawa, Ontario K1P 5P7, Canada.

Bjørke, Jan Terje (1988), "Kartinformasjon", Forelesningsnotat AID, Fylkeskartkontoret i Aust-Agder.

Bruegger, B. P. & Frank, A. U. (1990), "Hierarchical extensions of topographical datastructures". FIG cong. XIX, comm. 3, Helsinki Finland.

Burrough, P.A. (1986), "Principals of Geographical Information Systems for Land Resource Assessment", Oxford Science Publications, ISBN.

Cook, B.G. (1983), "An introduction to the design of geographic databases". In Proc. Workshop on databases in the natural sciences, CSIRO Division og computing Research 7.-9. September 1983, pp. 175-186, Cunningham Laboratory, Brisbane, Queensland.

Chrisman, N. R. (1990). "Deficiencies of sheet and tiles: building sheets databases". Int. Journal of Geographical Information Systems, no. 2.

Dale, P. F. & McLaughlin, J. D. (1988). "Land Information management. An introduction with special reference to cadastral problems in Third World countries", ISBN 0-19-858404-9, Claredon Press, Walton Street, Oxford, GB.

Einbu, John, (1987), "Geodatabaser", Tapir Trondheim.

Frank, Andrew U., "Requirements for a database Management System for a GIS", PE and RS, No. 11, 1988.

Goodchild, M. F. & Kemp, K. K. (1990). "National Centre for Geographic Information and Analysis. Core Curriculum: Introduction to GIS. Technical issues in GIS. Application Issues in GIS." University of California, Santa Barbara, USA.

Henriksen, Oskar (1988), "Begreper i digital kartproduskjon og et forslag til ajourholdstrategi", VIAK A/S (1988).

Ibbs, T.F., og Stevens, J. (1988), "Quadtree storage of vektor data", Int. J. Geographical Systems, nr. 1, 1988.

Johnson, Carol A., Detenbeck, Naomi E., Bonde, John P. and Biemi, Gerald J., "Geographical Information Systems for Cumulative Impact Assessment", PE and RS, no. 11, 1988.

Kroenke, D. (1977), "Database processing: fundamental modelling, applications", Science Research Associates, Chicago.

Lindblad. J. H. (1991). "Geographical Object-Orientation for Future Years". The University of Trondheim Norway.

Maguire, D. J. & Goodchild, M. F. & Rhind, D. W. (1991). "Geographic Information Systems". ISBN 0-582-05661-6, Lougman Scientific & Technical, Burat Hill, Harlow, Essex CM20 2JE, England.

Martin, J. (1983), "Managing the data-base environment", Prentice-Hall, Englewood Cliffs, N.J.

Moody, H.G. (1986), "Optical storage: mass storage with mass appeal?", Information Strategy, Vol 2, No. 4, pp 44.

Oxborrow, E. (1986), "Databases and database systems", Chartwell-Bratt. Lund-

Palmer, D. (1984), "A land information network for New Brunswick", Technical report no. 111, Departement of surveying Engeneering, University of New Brunswick, Frederiction, NB.

Parker, H. Dennison, "The Unique Qualities of a Geographic Information System: A Commentary", PE and RS, No 11, 1988.

Rhind, D.W. og Green, N.P.A. (1988), "Design of a geographical information system for a heterogeneous scientific community", Inst. J. Geographical Information Systems, No. 2 1988.

Robb, Margareth (1986), "Raster Geographic Information Systems, State of the Art", Notat, Norsk Regnesentral.

Vernoerd, Willy (1988), "Hva er egentlig relasjonsdatabaser", Datatid nr. 6/88.

CHAPTER 9

Aronoff, S. (1989), "Geographic Information Systems: A Management Perspective". ISBN 0-9218404-00-8, WDL Publications, P.O. Box 585, Station B, Ottawa, Ontario K1P 5P7, Canada.

Jan Terje Bjørke (1987), "Grunnleggende datastruktur og deres kartgeografiske anvendelse", kart og plan nr. 5.

Berg, Joseph (1987), "Fundamental operation in computer-assisted map analysis", Int. J. Geographical Information Systems, No. 2, 1987.

Berry, J. (1987). "Fundamental operations in computer-assisted map analyzing". Int. Journal of Geographical Information Systems, no. 2

Bobbe, Thomas (1987), "An application of a geographic information system to the timber sale planning process in the Tongass National Forest-Ketzhikna area", GIS '87, San Francisco.

Burrough, P.A. (1986), "Principals of Geographical Information Systems for Land Resource Assessment", Oxford Science Publications.

Chrisman, N. R. (1990). "Deficiencies of sheet and tiles: building sheets databases". Int. Journal of Geographical Information Systems, no. 2.

Dale, Petre, McLaughin, John (1988), "Land information Management", Claredon Press, Oxford.

Dale, P. F. & McLaughlin, J. D. (1988). "Land Information management. An introduction with special reference to cadastral problems in Third World countries", ISBN 0-19-858404-9, Clardon Press, Walton street, Oxford GB.

Dangermond, Jack, m.fl. (1987), "Network Analysis in Geographic Information Systems", PE and RS, No. 10,1987.

ESRI (1990). "Understanding GIS. The Arc/Info Method". ESRI, Redlands California, USA.

Goodchild, M. F. & Kemp, K. K. (1990). "National Centre for Geographic Information and Analysis. Core Curriculum: Introduction to GIS. Technical issues in GIS. Application Issues in GIS." University of California, Santa Barbara, USA.

Jenkson, S.K. and Dominique, J.O., "Extracting Topographic Structure from Digital Evaluation Data for Geographic Information System Analysis", PE and RS, No. 11, 1988.

Justusson, Bo, Olsson, Stig (1987), "SCB-rutkartor på persondata", Svensk Lantmäteritidsskrift nr. 4-5, 1987.

Maguire, D. J. & Goodchild, M. F. & Rhind, D. W. (1991). "Geographic Information Systems". ISBN 0-582-05661-6, Lougman Scientific & Technical, Burat Hill, Harlow, Essex CM20 2JE, England.

Robb, M. (1986). "Raster geographic information systems. State of the art". Norwegian Computer Centre.

Robb, Margaret (1986), "The Application of Expert System Techniques to Geographic Information Systems", Notat 16/89, Norsk Regnesentral.

Robinson, Vincent (1987), "Expert Systems for Geographic Information Systems", PE og RS nr. 10, 1987.

Smith, S.M., m.fl. (1987), "Agricultural field management with micro-computer based GIS and image analysis systems", GIS '87, San Francisco.

Strand, G-H. (1991), "Linear Combination Models in Geographical Information Systems". NITO-kurs, Norway.

Strand, G. H. (1991), "Sentrale teknikker i geografiske informasjonssystemer". NITO-kurs, Norway.

Strand, Geir Harald og Kvenild, Lars (1984), "Supermap, Software for tematisk kartografili", notat nr. 43, Geografisk Institutt, Universitetet Trondheim.

Østman, Anders (1987), "Metod for analys av legsbunden information", Svensk Lantmäteritidsskrift nr. 4-5, 1987.

CHAPTER 10

Baudouin, Axel, Anker, Peder (1984), "Kartpresentasjon og EDB-assistert kartografi" Publ. nr. 744, Norsk Regnesentral.

Bjørke, Jan Terje (1988), "Kartinformasjon", Forelesningsnotat AID, Fylkeskartkontoret i Aust-Agder.

Goodchild, M. F. & Kemp, K. K. (1990). "National Centre for Geographic Information and Analysis. Core Curriculum: Introduction to GIS. Technical issues in GIS. Application Issues in GIS." University of California, Santa Barbara, USA.

Maguire, D. J. & Goodchild, M. F. & Rhind, D. W. (1991). "Geographic Information Systems". ISBN 0-582-05661-6, Lougman Scientific & Technical, Burat Hill, Harlow, Essex CM20 2JE, England.

Palm, C. (1990), "Karttecken – kartsymbol". Nordisk sommerkurs, NKTF, Norway.

CHAPTER 11

Aas, A. G. (1991). "Informasjonsanalyse. Et strategisk verktøy for informasjonshåndtering". VIAK A/S, Norway

Adam, Jørgen (1983), "Bakgrunn og målsetting. Samfunnets produksjon av kart og geodata. Metoden for kvantifisering av kart og geodata. Metoden for kvantifisering av nytteverdi". KVANTIF, VIAK A/S, 1983.

Alexander, David og Stent, Bill (1989), "GIS-making it low on costs and high on benefits", ESRI User Conference, 1989.

Angus-Leppan, P. (1983), "Economics costs and benefits of land information", Proceedings of the XVII Congress of the International Federation of Surveyors, Vol. 3, Sophia.

Bernhardsen, T. (1986), "A cost-benefit study of a GIS-methodology and result". Paper presented to the FIG-International Congress, Toronto, June.

Bernstein, J. (1985), "The costs of land information systems". Paper presented to the World Bank Seminar on Land Information systems, St. Michales, Md.

Bie, Stein (1987), "Om fremtiden", Norsk Regnesentral.

Burrough, P.A. (1986), "Principals of Geographical Information Systems for Land Resource Assessment", Oxford Science Publications.

Dale, Petre, McLaughin, John (1988), "Land information Management", Claredon Press, Oxford.

DeMan, Eric (1988), "Establishing a geographical information system in relation to its use. A process of strategic choices", Inst. J. Geographical Information Systems, nr. 3, 1988.

Dickinson, Holley og Calcins, Hugh (1988), "The Economic evaluation of implementing a GIS", Int. J. Geographical Information Systems, nr. 4, 1988.

Gorman, M. (1984), "Managing database: four critical factors", Wellesly, Mass.: QED Publications.

Hansen, D. (1984), "An overview of the organization, costs and benefits of computer assisted mapping and records activity systems", Computers, Environments and Urban Systems Vol. 9, No. 2/3.

Hamiltong, A, er al. (1985), "Unit costs principles and their application to property mapping in New Brunswick, The Canadian Surveyor Vol. 39/1, pp. 11-22.

Harrison, J. and Dangermond, J. (1989). "Five tracks to GIS development and implementation". ESRI, Redlands California, USA.

Havt, A.L. (1986), "The Politics of Land Information Systems", FIG-congress 18, Toronto 1986.

Joofe, Bruce (1987), "Evaluating and selecting a GIS system", Proceedings, GIS '87-San Francisco.

Jones, Ken (1988), "Innføring i kartdatateknikk i mindre kommuner", Norsk institutt for by- og regionalforskning.

Kylen, B. & Hekland, J. (1990). "Economics of geographic information. Organizational Impact of Technological Changes in the Road GIS case". Nordic KVANTIF, Secretary VIAK, Arendal Norway.

Maguire, D. J. & Goodchild, M. F. & Rhind, D. W. (1991). "Geographic Information Systems". ISBN 0-582-05661-6, Lougman Scientific & Technical, Burat Hill, Harlow, Essex CM20 2JE, England.

Man, W.H. de, ed (1984), "Conceptual framework and guidelines for establishing geographic information systems", General Information Programme and UNISIST, UNESCO, Paris.

Nordic KVANTIF (1987), "Community benefit of digital spatial information-digital maps data bases, economics and user experiences in North America (NORDISK KVANTIF)", Prosjektrappert nr. 1, 2, 3, VIAK A/S.

Nordic KVANTIF (1987). "Community Benefit of Digital Spatial Information. Final report". Viak, Arendal, Norway.

Nordic KVANTIF (1987), "Samhällsnyttan av digital lägsbunden information-slutrapport (NORDISK KVANTIF)", Prosjektrappert nr. 5, VIAK A/S.

Scholten, Herke og Padding, Paul (1988), "Working with GIS in policy environment", ESRI-User Conference, München, 1988.

Statskonsult (1990). "Kost-nytteanalyse av IT-prosjekter". Direktoratet for statsforvaltningsutvikling, Norway.

CHAPTER 12

Falloux, F. (1992), "Environmental Information, one of the Keys to Success". From a coming book.

Drummond, J. & Stefanovic, P. (1986). "Transfer of high technology to developing countries". Auto Carto, London.

Falloux, F. (1989). "Land Information and Remote Sensing for Renewable Resource Management in Sub-Saharan Africa". The World Bank, USA.

van Genderen, J. L. (1991). "Guidelines for education and training in environmental information systems in Sub-Saharan Africa: Some key issues". ITC, The Netherlands.

CHAPTER 13

Kyrkjeide, A. & Sandvik R. (1990). "Elektroniske sjøkart". Norges Sjøkartverk, Norway.

Norwegian Hydrographic Service (1991), "The Seatrans Project, 1989 - 90, Final Report"

CHAPTER 14

Granheim, O. (1990), "GIS-prosjektet i Oslo kommune. Kostnad-nytte". Kart og Plan nr. 3 1990, Norway.

Jelhma, J. (1987), "A GIS for National Planning", ESRI User onference, München, 1987.

McDonnels, Douglas Information Systems AB (1986), "Stockholm Central-Spårprojektering på CAD".

Nordic KVANTIF (1987), "Community benefit of digital spatial information-digital maps data bases, economics and user experiences in North America (NORDISK KVANTIF)", Prosjektrapport nr. 1, 2, 3, VIAK A/S.

Nordic KVANTIF (1987), "Community benefit of digital spatial information-digital maps databases, economics and user experience in North America", VIAK A/S

INDEX